Gene Prediction: Applying Ontology and Machine Learning (Volume I)

Gene Prediction: Applying Ontology and Machine Learning (Volume I)

Casper Harvey

Larsen & Keller
www.larsen-keller.com

Gene Prediction: Applying Ontology and Machine Learning (Volume I)
Casper Harvey
ISBN: 979-8-88836-052-1 (Hardback)

Larsen & Keller
Published by Larsen and Keller Education,
5 Penn Plaza,
19th Floor,
New York, NY 10001, USA

Cataloging-in-Publication Data

Gene prediction : applying ontology and machine
learning (Volume I) / Casper Harvey.
p. cm.
Includes bibliographical references and index.
ISBN 979-8-88836-052-1
1. Genomics--Data processing. 2. Bioinformatics. 3. Genomics. 4. Data mining.
5. Genes. 6. Molecular genetics. I. Harvey, Casper.
QH447 .G465 2023
572.86--dc23

For more information regarding Larsen and Keller Education and its products, please visit the publisher's website www.larsen-keller.com

Table of Contents

Preface

Every book is a source of knowledge and this one is no exception. The idea that led to the conceptualization of this book was the fact that the world is advancing rapidly; which makes it crucial to document the progress in every field. I am aware that a lot of data is already available, yet, there is a lot more to learn. Hence, I accepted the responsibility of editing this book and contributing my knowledge to the community.

Gene prediction refers to the process of identifying the regions of genomic DNA that encodes genes using computational methods. It is an important part of bioinformatics. Gene prediction is the first step for annotating large and contiguous sequences. It aids in identifying the essential elements of the genome including functional genes, intron, splicing sites, exon, and regulatory sites. It is also used in describing the individual genes based on their functions. Protein function prediction is an important part of genome annotation. Lately, high-throughput sequencing technologies have led to development of prediction methods. Gene ontology (GO) is one of the databases that are available for identifying the functional properties of proteins. Research in this domain is now focused on efficiently predicting the GO terms. Researches are ongoing on the use of machine learning algorithms for functional prediction as these algorithms use rule-based approaches to integrate large amounts of heterogeneous data and detect patterns. mSplicer, mGene, and CONTRAST are methods that use machine learning techniques for gene prediction. Gene prediction methods are widely used in fields like structural genomics, functional genomics, and genome studies. This book traces the progress of gene prediction and the application of ontology and machine learning. It is appropriate for students seeking detailed information in this area of study as well as for experts.

While editing this book, I had multiple visions for it. Then I finally narrowed down to make every chapter a sole standing text explaining a particular topic, so that they can be used independently. However, the umbrella subject sinews them into a common theme. This makes the book a unique platform of knowledge.

I would like to give the major credit of this book to the experts from every corner of the world, who took the time to share their expertise with us. Also, I owe the completion of this book to the never-ending support of my family, who supported me throughout the project.

Casper Harvey

1

Statistical Learning and Genomic Selection: Basic Concepts

1.1 Data as a Powerful Weapon

Thanks to advances in digital technologies like electronic devices and networks, it is possible to automatize and digitalize many jobs, processes, and services, which are generating huge quantities of data. These "big data" are transmitted, collected, aggregated, and analyzed to deliver deep insights into processes and human behavior. For this reason, data are called the new oil, since "data are to this century what oil was for the last century"—that is, a driver for change, growth, and success. While statistical and machine learning algorithms extract information from raw data, information can be used to create knowledge, knowledge leads to understanding, and understanding leads to wisdom (Sejnowski 2018). We have the tools and expertise to collect data from diverse sources and in any format, which is the cornerstone of a modern data strategy that can unleash the power of artificial intelligence. Every single day we are creating around 2.5 quintillion bytes of data (McKinsey Global Institute 2016). This means that almost 90% of the data in the world has been generated over the last 2 years. This unprecedented capacity to generate data has increased connectivity and global data flows through numerous sources like tweets, YouTube, blogs, sensors, internet, Google, emails, pictures, etc. For example, Google processes more than 40,000 searches every second (and 3.5 billion searches per day), 456,000 tweets are sent, and 4,146,600 YouTube videos are watched per minute, and every minute, 154,200 Skype calls are made, 156 million emails are sent, 16 million text messages are written, etc. In other words, the amount of data is becoming bigger and bigger (big data) day by day in terms of volume, velocity, variety, veracity, and "value."

The nature of international trade is being radically transformed by global data flows, which are creating new opportunities for businesses to participate in the global economy. The following are some ways that these data flows are transforming international trade: (a) businesses can use the internet (i.e., digital platforms) to export goods; (b) services can be purchased and consumed online; (c) data collection and analysis are allowing new services (often also provided online) to add value to exported goods; and (d) global data flows underpin global value chains, creating new opportunities for participation.

According to estimates, by 2020, 15–20% of the gross domestic product (GDP) of countries all over the world will be based on data flows. Companies that adopt big data practices are expected to increase their productivity by 5–10% compared to companies that do not, and big data practices could add 1.9% to Europe's GDP between 2014 and 2020. According to McKinsey Global Institute (2016) estimates, big data could generate an additional $3 trillion in value every year in some industries. Of this, $1.3 trillion would benefit the United States. Although these benefits do not directly affect the GDP or people's personal income, they indirectly help to improve the quality of life.

But once we have collected a large amount of data, the next question is: how to make sense of it? A lot of businesses and people are collecting data, but few are extracting useful knowledge from them. As mentioned above, nowadays there have been significant advances for measuring and collecting data; however, obtaining knowledge from (making sense of) these collected data is still very hard and challenging, since there are few people in the market with the expertise and training needed to extract knowledge from huge amounts of data. For this reason, to make sense of data, new disciplines were recently created, such as data science (the commercial name of statistics), business intelligence and analytics, that use a combination of statistics, computing, machine learning, and business, among other domains, to analyze and discover useful patterns to improve the decision-making process. In general, these tools help to (1) rationalize decisions and the decision-making process, (2) minimize errors and deal with a range of scenarios, (3) get a better understanding of the situation at hand (due to the availability of historical data), (4) assess alternative solutions (based on key performance indicators), and (5) map them against the best possible outcome (benchmarking), which helps make decision-making more agile. For this reason, data analysis has been called "the sexiest job of the twenty-first century." For example, big data jobs in the United States are estimated to be 500,000. But the McKinsey Global Institute (2016) estimates that there is still a shortage of between 140,000 and 190,000 workers with a background in statistics, computer engineering, and other applied fields, and that 1.5 million managers are needed to evaluate and make decisions on big data. This means that 50–60% of the required staff was lacking in the year 2018 in the United States alone (Dean 2018).

Some areas and cases where statistical and machine learning techniques have been successfully applied to create knowledge and make sense of data are given next. For example, in astronomy these techniques have been used for the classification of exoplanets using thousands of images taken of these celestial bodies. Banks use them to decide if a client will be granted credit or not, using as predictors many socioeconomic variables they ask of clients. Banks also use them to detect credit card fraud. On the internet, they are used to classify emails as spam or ham (not spam) based on previous emails and the text they contain. In genomic selection, they are used to predict grain yield (or another trait) of non-phenotyped plants using information, including thousands of markers and environmental data. In Google, they are used for recommending books, movies, products, etc., using previous data on the characteristics (age, gender, location, etc.) of the people who have used these services. They are also used in self-driving cars, that is, cars capable of sensing their environment and moving with little or no human input, based on thousands of images and information from sensors that perceive their surroundings. These examples give more evidence that the appropriate use of data is a powerful weapon for getting knowledge of the target population.

However, as is true of any new technology, this data-related technology can be used against society. One example is the Facebook–Cambridge Analytica data scandal that was a major political scandal in early 2018 when it was revealed that Cambridge Analytica had harvested personal data from the Facebook profiles of millions of people without their consent and used them for political purposes. This scandal was described as a watershed moment in the public understanding of personal data and precipitated a massive fall in Facebook's stock price and calls for tighter regulation of tech companies' use of data. For these reasons, some experts believe that governments have a responsibility to create and enforce rules on data privacy, since data are a powerful weapon, and weapons should be controlled, and because privacy is a fundamental human right. All these are very important to avoid the weaponization of data against people and society.

To take advantage of the massive data collected in Genomic Selection (GS) and many other domains, it is really important to train people in statistical machine learning methods and related areas to perform precise prediction, extract useful knowledge, and find hidden data patterns. This means that experts in statistical machine learning methods should be able to identify the statistical or machine learning method that is most relevant to a given problem, since there is no universal method that works well for all data sets, cleans the original data, implements these methods in statistical machine learning software, and interprets the output of statistical machine learning methods correctly to translate the big data collected into insights and operational value quickly and accurately.

1.2 Genomic Selection

Plant breeding is a key scientific area for increasing the food production required to feed the people of our planet. The key step in plant breeding is selection, and conventional breeding is based on phenotypic selection. Breeders choose good offspring using their experience and the observed phenotypes of crops, so as to achieve genetic improvement of target traits (Wang et al. 2018). Thanks to this area (and related areas of science), the genetic gain nowadays has reached a near-linear increase of 1% in grain yield yearly (Oury et al. 2012; Fischer et al. 2014). However, a linear increase of at least 2% is needed to cope with the 2% yearly increase in the world population, which relies heavily on wheat products as a source of food (FAO 2011). For this reason, genomic selection (GS) is now being implemented in many plant breeding programs around the world. GS consists of genotyping (markers) and phenotyping individuals in the reference (training) population and, with the help of statistical machine learning models, predicting the phenotypes or breeding values of the candidates for selection in the testing (evaluation) population that were only genotyped. GS is revolutionizing plant breeding because it is not limited to traits determined by a few major genes and allows using a statistical machine learning model to establish the associations between markers and phenotypes and also to make predictions of non-phenotyped individuals that help make a more comprehensive and reliable selection of candidate individuals. In this way, it is essential for accelerating genetic progress in crop breeding (Montesinos-López et al. 2019).

1.2.1 Concepts of Genomic Selection

The development of different molecular marker systems that started in the 1980s drastically increased the total number of polymorphic markers available to breeders and molecular biologists in general. The single nucleotide polymorphism (SNP) that has been intensively used in QTL discovery is perhaps the most popular high-throughput genotyping system (Crossa et al. 2017). Initially, by applying marker-assisted selection (MAS), molecular markers were integrated with traditional phenotypic selection. In the context of simple traits, MAS consists of selecting individuals with QTL-associated markers with major effects; markers not significantly associated with a trait are not used (Crossa et al. 2017). However, after many attempts to improve complex quantitative traits by using QTL-associated markers, there is not enough evidence that this method really can be helpful in practical breeding programs due to the difficulty of finding the same QTL across multiple environments (due to QTL × environment interaction) or in different genetic backgrounds (Bernardo 2016). Due to this difficulty of the MAS approach, in the early 2000s, an approach called association mapping appeared with the purpose of overcoming the insufficient power of linkage analysis, thus facilitating the detection of marker–trait associations in non-biparental populations and fine-mapping chromosome segments with high recombination rates (Crossa et al. 2017). However, even the fine-mapping approach was unable to increase the power to detect rare variants that may be associated with economically important traits.

For this reason, Meuwissen et al. (2001) proposed the GS methodology (that was initially used in animal science), which is different from association mapping and QTL analysis, since GS simultaneously uses all the molecular markers available in a training data set for building a prediction model; then, with the output of the trained model, predictions are performed for new candidate individuals not included in the training data set, but only if genotypic information is available for those candidate individuals. This means that the goal of GS is to predict breeding and/or genetic values. Because GS is implemented in a two-stage process, to successfully implement it, the data must be divided into a training (TRN) and a testing (TST) set, as can be observed in Fig. 1.1. The training set is used in the first stage, while the testing set

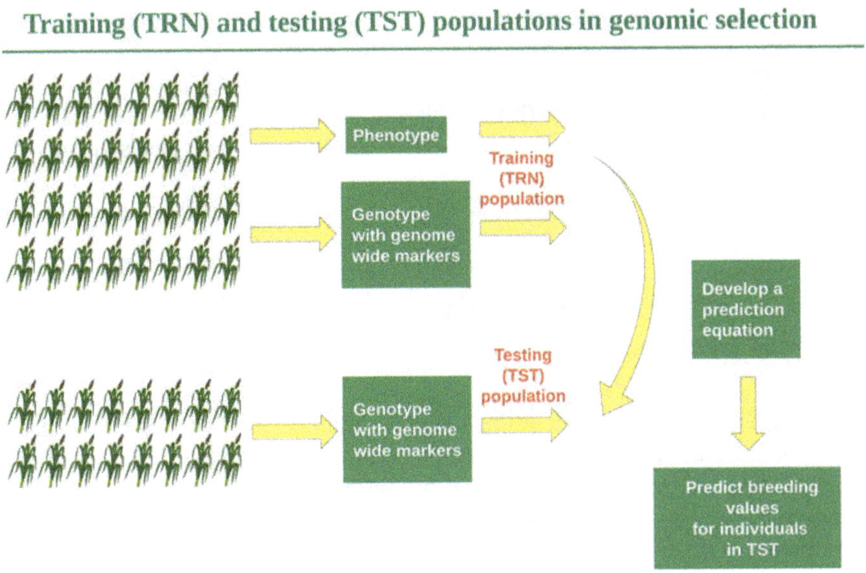

Fig. 1.1 Schematic representation of TRN and TST sets required for implementing GS (Crossa et al. 2017)

is used in the second stage. The main characteristics of the training set are (a) it combines molecular (independent variables) and phenotypic (dependent variables) data and (b) it contains enough observations (lines) and predictors (molecular data) to be able to train a statistical machine learning model with high generalized power (able to predict data not used in the training process) to predict new lines. The main characteristic of the testing set is that it only contains genotypic data (markers) for a sample of observations (lines) and the goal is to predict the phenotypic or breeding values of lines that have been genotyped but not phenotyped.

The two basic populations in a GS program are shown in Fig. 1.1: the training (TRN) data whose genotype and phenotype are known and the testing (TST) data whose phenotypic values are to be predicted using their genotypic information. GS substitutes phenotyping for a few selection cycles. Some advantages of GS over traditional (phenotypic) selection are that it: (a) reduces costs, in part by saving the resources required for extensive phenotyping, (b) saves time needed for variety development by reducing the cycle length, (c) has the ability to substantially increase the selection intensity, thus providing scenarios for capturing greater gain per unit time, (d) makes it possible to select traits that are very difficult to measure, and (e) can improve the accuracy of the selection process. Of course, successful implementation of GS depends strongly on the quality of the training and testing sets.

GS has great potential for quickly improving complex traits with low heritability, as well as significantly reducing the cost of line and hybrid development. Certainly, GS could also be employed for simple traits with higher heritability than complex traits, and high genomic prediction (GP) accuracy is expected. Application of GS in plant breeding could be limited by some of the following factors: (i) genotyping cost, (ii) unclear guidelines about where in the breeding program GS could be efficiently applied (Crossa et al. 2017), (iii) insufficient number of lines or animals in the reference (training) population, (iv) insufficient number of SNPs in the panels, and (v) the reference population contains very heterogeneous individuals (plants or animals).

1.2.2 Why Is Statistical Machine Learning a Key Element of Genomic Selection?

GS is challenging and very interesting because it aims to improve crop productivity to satisfy humankind's need for food. Addressing the current challenges to increase crop productivity by improving the genetic makeup of plants and avoiding plant diseases is not new, but it is of paramount importance today to be able to increase crop productivity around the world without the need to increase the arable land. Statistical machine learning methods can help improve GS methodology, since they are able to make computers learn patterns that could be used for analysis, interpretation, prediction, and decision-making. These methods learn the relationships between the predictors and the target output using statistical and mathematical models that are implemented using computational tools to be able to predict (or explain) one or more dependent variables based on one or more independent variables in an efficient manner. However, to do this successfully, many real-world problems are only approximated using the statistical machine learning tools, by evaluating probabilistic distributions, and the decisions made using these models are supported by indicators like confidence intervals. However, the creation of models using probability distributions and indicators for evaluating prediction (or association) performance is a field of statistical machine learning, which is a branch of artificial intelligence, understanding as statistical machine learning the application of statistical methods to identify patterns in data using computers, but giving computers the ability to learn without being explicitly programmed (Samuel 1959). However, artificial intelligence is the field of science that creates machines or devices that can mimic intelligent behaviors.

As mentioned above, statistical machine learning allows learning the relationship between two types of information that are assumed to be related. Then one part of the information (input or independent variables) can be used to predict the information lacking (output or dependent variables) in the other using the learned relationship. The information we want to predict is defined as the response variable (y), while the information we use as input are the predictor variables ($\boldsymbol{X}$). Thanks to the continuous reduction in the cost of genotyping, GS nowadays is implemented in many crops around the world, which has caused the accumulation of large amounts of biological data that can be used for prediction of non-phenotyped plants and animals. However, GS implementation is still challenging, since the quality of the data (phenotypic and genotypic) needs to be improved. Many times the genotypic information available is not enough to make high-quality predictions of the target trait, since the information available has a lot of noise. Also, since there is no universal best prediction model that can be used under all circumstances, a good understanding of statistical machine learning models is required to increase the efficiency of the selection process of the best candidate individuals with GS early in time. This is very important because one of the key components of genomic selection is the use of statistical machine learning models for the prediction of non-phenotyped individuals. For this reason, statistical machine learning tools have the power to help increase the potential of GS if more powerful statistical machine learning methods are developed, if the existing methods can deal with larger data sets, and if these methods can be automatized to perform the prediction process with only a limited knowledge of the subject.

For these reasons, statistical machine learning tools promise considerable benefits for GS and agriculture through their contribution to productivity growth and gain in the genetic makeup of plants and animals without the need to increase the arable land. At the same time, with the help of statistical machine learning tools, GS is deeply impacting the traditional way of selecting candidate individuals in plant and animal breeding. Since GS can reduce by at least half the time needed to select candidate individuals, it has been implemented in many crops around the globe and is radically changing the traditional way of developing new varieties and animals all over the world.

Although GS is not the dominant paradigm for developing new plants and animals, it has the potential to transform the way they are developed due to the following facts: (a) the massive amounts of data being generated in plant breeding programs are now available to train the statistical machine learning methods, (b) new technologies such as sensors, satellite technology, and robotics allow scientists to generate not only genomic data but also phenotypic data that can capture a lot of environmental and phenotypic information that can be used in the modeling process to increase the performance of statistical machine learning methods, (c) increased computational power now allows complex statistical machine learning models with larger data sets to be implemented in less time, and (d) there is now greater availability of user-friendly statistical machine learning software for implementing a great variety of statistical machine learning models.

However, there are still limitations for the successful implementation of GS with the help of statistical machine learning methods because much human effort is required to collect a good training data set for supervised learning. Although nowadays it is possible to measure a lot of independent variables (markers, environmental variables) due to the fact that the training set should be measured in real-world experiments conducted in different environments and locations, this is expensive and subject to nature's random variability. This means that GS data are hampered by issues such as multicollinearity among markers (adjacent markers are highly correlated) and by a problem that consists of having a small number of observations and a large number of independent variables (commonly known as "large *p* small *n*"), which poses a statistical challenge. For this reason, obtaining data sets that are large and comprehensive enough to be used for training—for example, creating or obtaining sufficient plant trial data to predict yield, plant height, grain quality, and presence or absence of disease outcomes more accurately—is also often challenging.

Another challenge is that of building statistical machine learning techniques that are able to generalize the unseen data, since statistical machine learning methods continue to have difficulty carrying their experiences from one set of circumstances to another. This is known as transfer learning, and it focuses on storing knowledge gained when training a particular machine learning algorithm and then using this stored knowledge for solving another related problem. In other words, transfer learning is still very challenging and occurs when a statistical machine learning model is trained to accomplish a certain task and then quickly apply that learning exercise to a different activity.

Another disadvantage is that even though today there are many software programs for implementing statistical machine learning tools for GS, the computational resources required for learning from moderate to large data sets are very expensive and most of the time it is not possible to implement them in commonly used computers, since servers with many cores and considerable computational resources are required. However, the rapid increase in computational power will change this situation in the coming years.

1.3 Modeling Basics

1.3.1 What Is a Statistical Machine Learning Model?

A model is a simplified description, using mathematical tools, of the processes we think that give rise to the observations in a set of data. A model is deterministic if it explains (completely) the dependent variables based on the independent ones. In many real-world scenarios, this is not possible. Instead, statistical (or stochastic) models try to approximate exact solutions by evaluating probabilistic distributions. For this reason, a statistical model is expressed by an equation composed of a *systematic* (deterministic) and a *random part* (Stroup 2012) as given in the next equation:

$$y_i = f(\boldsymbol{x}_i) + \epsilon_i, \text{for } i = 1, 2, \ldots, n, \tag{1.1}$$

where y_i represents the response variable in individual i and $f(\boldsymbol{x}_i)$ is the systematic part of the model because it is *determined* by the explanatory variables (predictors). For these reasons, the systematic part of the statistical learning model is also called the deterministic part of the model, which gives rise to an unknown mathematical function (f) of $\boldsymbol{x}_i = x_{i1}, \ldots, x_{ip}$ not subject to random variability. ϵ_i is the ith random element (error term) which is independent of $\boldsymbol{x}_i$ and has mean zero. The ϵ_i term tells us that observations are assumed to vary at random about their mean, and it also defines the uniqueness of each individual. In theory (at least in some philosophical domains), if we know the mechanism that gives rise to the uniqueness of each individual, we can write a completely deterministic model. However, this is rarely possible because we use probability distributions to characterize the observations measured in the individuals. Most of the time, the error term (ϵ_i) is assumed to follow a normal distribution with mean zero and variance σ^2 (Stroup 2012).

As given in Eq. (1.1), the f function that gives rise to the systematic part of a statistical learning model is not restricted to a unique input variable, but can be a function of many, or even thousands, of input variables. In general, the set of approaches for estimating f is called statistical learning (James et al. 2013). Also, the functions that f can take are very broad due to the huge variety of phenomena we want to predict and due to the fact that there is no universally superior f that can be used for all processes. For this reason, to be able to perform good predictions out of sample data, many times we need to fit many models and then choose the one most likely to succeed with the help of cross-validation techniques. However, due to the fact that models are only a simplified picture of the true complex process that gives rise to the data at hand, many times it is very hard to find a good candidate model. For this reason, statistical machine learning provides a catalog of different models and algorithms from which we try to find the one that best fits our data, since there is no universally best model and because there is evidence that a set of assumptions that works well in one domain may work poorly in another—this is called the *no free lunch theorem* by Wolpert (1996). All these are in agreement with the famous aphorism, "all models are wrong, but some are useful," attributed to the British statistician George Box (October 18, 1919–March 28, 2013) who first mentioned this aphorism in his paper "Science and Statistics" published in the *Journal of the American Statistical Association* (Box 1976). As a result of the *no free lunch theorem*, we need to evaluate many models, algorithms, and sets of hyperparameters to find the best model in terms of prediction performance, speed of implementation, and degree of complexity. This book is concerned precisely with the appropriate combination of data, models, and algorithms needed to reach the best possible prediction performance.

1.3.2 The Two Cultures of Model Building: Prediction Versus Inference

The term "two cultures" in statistical model building was coined by Breiman (2001) to explain the difference between the two goals for estimating f in Eq. (1.1): prediction and inference. These definitions are provided in order to clarify the distinct scientific goals that follow inference and empirical predictions, respectively. A clear understanding and distinction between these two approaches is essential for the progress of scientific knowledge. Inference and predictive modeling reflect the process of using data and statistical (or data mining) methods for inferring or predicting, respectively. The term modeling is intentionally chosen over model to highlight the entire process involved, from goal definition, study design, and data collection to scientific use (Breiman 2001).

Prediction

The prediction approach can be defined as the process of applying a statistical machine learning model or algorithm to data for the purpose of predicting new or future observations. For example, in plant breeding a set of inputs (marker information) and the outcome Y (disease resistance: yes or no) are available for some individuals, but for others only marker information is available. In this case, marker information can be used as a predictor and the disease status should be used as the response variable. When scientists are interested in predicting new plants not used to train the model, they simply want an accurate model to predict the response using the predictors. However, when scientists are interested in understanding the relationship between each individual predictor (marker) and the response variable, what they really want is a model for inference. Another example is when forest scientists are interested in developing models to predict the number of fire hotspots from an accumulated fuel dryness index, by vegetation type and region. In this context, it is obvious that scientists are interested in future predictions to improve decision-making in forest fire management. Another example is when an agro-industrial engineer is interested in developing an automated system for classifying mango species based on hundreds of mango images taken with digital cameras, mobile phones, etc. Here again it is clear that the best approach to build this system should be based on prediction modeling since the objective is the prediction of new mango species, not any of those used for training the model.

Inference

Many areas of science are devoted mainly to testing causal theories. Under this framework, scientists are interested in testing the validity of a theoretical causal relationship between the causal variables (X; underlying factors) and the measured variable (Y) using statistical machine learning models and collected data to test the causal hypotheses. The type of statistical machine learning models used for testing causal hypotheses are usually association-based models applied to observational data (Shmueli 2012). For example, regression models are one type of association-based models used for testing causal hypotheses. This practice is justified by the theory itself, which assumes the causality. In this context, the role of the theory is very strong and the reliance on data and statistical modeling is strictly through the lens of the theoretical model. The theory–data relationship varies in different fields. While the social sciences are very theory-heavy, in areas such as bioinformatics and natural language processing, the emphasis on a causal theory is much weaker. Hence, given this reality, Shmueli (2012) defined *explaining* as causal explanation and *explanatory modeling* as the use of statistical models for testing causal explanations.

Next, we provide some great examples used for testing causal hypotheses: for example, between 1911 and 1912, Austrian physicist Victor Hess made a series of ten balloon ascents to study why metal plates tend to charge spontaneously. At that time, it was assumed that the cause was the presence, in small quantities, of radioactive materials in rocks. If this is the case, as one moves away from the ground, the tendency of metal plates to be charged should decrease. Hess brought with him three electroscopes, which are instruments composed basically of two metal plates enclosed in a glass sphere. When charging the plates, they separated one from the other. Hess observed that from a certain height, the three electroscopes tended to be charged to a greater extent. On August 7, 1912, together with a flight commander and a meteorologist, he made a 6-hour flight in which he ascended to more than 5000 m in height (Schuster 2014). Hess published his results in 1913, where he presented his conclusion that the cause of the charge of the electroscopes was radiation of cosmic origin that penetrates the atmosphere from above. The discovery of this cosmic radiation, for which Hess received the Nobel Prize in 1936, opened a new window for the study of the universe (Schuster 2014). It was in 1925 when American physicist Robert Andrew Millikan introduced the term “cosmic rays” to describe this radiation, and what it was made of was still unknown. In this example, it is clear that the goal of the analysis was association.

No one suspected that tobacco was a cause of lung tumors until the final decade of the nineteenth century. In 1898, Hermann Rottmann (a medical student) in Würzburg proposed that tobacco dust—not smoke—might be causing the elevated incidence of lung tumors among German tobacco workers. This was a mistake corrected by Adler (1912) who proposed that smoking might be to blame for the growing incidence of pulmonary tumors. Lung cancer was still a very rare disease; so rare, in fact, that medical professors, when confronted with a case, sometimes told their students they might never see another. However, in the 1920s, surgeons were already faced with a greater incidence of lung cancer, and they began to get confused about its possible causes. In general, smoking was blamed, along with asphalt dust from recently paved roads, industrial air pollution, and the latent effects of poisonous gas exposure during World War I or the global influenza pandemic of 1918–1919. These and many other theories were presented as possible explanations for the increase in lung cancer, until evidence from multiple research sources made it clear that tobacco was the main culprit (Proctor 2012). Here again it is clear that the goal of the analysis should be related to inference.

1.3.3 Types of Statistical Machine Learning Models and Model Effects

1.3.3.1 Types of Statistical Machine Learning Models

Statistical machine learning models are most commonly classified as parametric models, semiparametric models, and nonparametric models. Next, we define each type of statistical machine learning models and provide examples that help to understand each one.

Parametric Model It is a type of statistical machine learning model in which all the predictors take predetermined forms with the response. Linear models (e.g., multiple regression: $y = \beta_1 x_1 + \beta_2 x_2 + \beta_3 x_3 + \epsilon$), generalized linear models [Poisson regression: $E(y|x) = \exp(\beta_1 x_1 + \beta_2 x_2 + \beta_3 x_3)$], and nonlinear models (nonlinear regression: $y = \beta_1 x_1 + \beta_2 x_2 + \beta_3 e^{\beta_4 x_3} + \epsilon$) are examples of parametric statistical machine learning models because we know the function that describes the relationship between the response and the explanatory variables. These models are very easy to interpret but very inflexible.

Nonparametric Model It is a type of statistical machine learning model in which none of the predictors take predetermined forms with the response but are constructed according to information derived from data. Two common statistical machine learning models are kernel regression and smoothing spline. Kernel regression estimates the conditional expectation of y at a given value x using a weighted filter on the data ($y = m(x) + \epsilon$, with $\widehat{m}(x_0) = \frac{\sum_{i=1}^{n} K\left(\frac{x_i - x_0}{h}\right) y_i}{\sum_{i=1}^{n} K\left(\frac{x_i - x_0}{h}\right)}$), where h is the bandwidth (this estimator of $m(x)$ is called the *Nadaraya–Watson (NW) kernel estimator*) and K is a kernel function. While smoothing splines minimize the sum of squared residuals plus a term which penalizes the roughness of the fit [$y = \beta_0 + \beta_1 x + \beta_2 x^2 + 3x^3 + \sum_{j=1}^{J} \beta_{1j} (x - \theta_j)_+^3$, where $(x - \theta_j)_+ = x - \theta_j$, $x > \theta_j$ and 0 otherwise], this model in brackets is a spline of degree 3 which is represented as a power series. These models are very difficult to interpret but are very flexible. Nonparametric statistical machine learning models differ from parametric models in that the shape of the functional relationships between the response (dependent) and the explanatory (independent) variables are not predetermined but can be adjusted to capture unusual or unexpected features of the data. Nonparametric statistical machine learning models can reduce modeling bias (the difference between estimated and true values) by imposing no specific model structure other than certain smoothness assumptions, and therefore they are particularly useful when we have little information or we want to be flexible about the underlying statistical machine learning model. In general, nonparametric statistical machine learning models are very flexible and are better at fitting the data than parametric statistical machine learning models. However, these models require larger samples than parametric statistical machine learning models because the data must supply the model structure as well as the model estimates.

Semiparametric Model It is a statistical machine learning model in which *part* of the predictors do not take predetermined forms while the other part takes known forms with the response. Some examples are (a) $y = \beta_1 x_1 + \beta_2 x_2 + \beta_3 x_3 + m(x) + \epsilon$ and (b) $y = \exp(\beta_1 x_1 + \beta_2 x_2 + \beta_3 x_3) + m(x) + \epsilon$. This means that semiparametric models are a mixture of parametric and nonparametric models.

When the relationship between the response and explanatory variables is known, parametric statistical machine learning models should be used. If the relationship is unknown and nonlinear, nonparametric statistical machine learning models should be used. When we know the relationship between the response and part of the explanatory variables, but do not know the relationship between the response and the other part of the explanatory variables, we should use semiparametric statistical machine learning models. Any application area that uses statistical machine learning analysis could potentially benefit from semi/nonparametric regression.

1.3.3.2 Model Effects

Many statistical machine learning models are expressed as models that incorporate *fixed effects*, which are parameters associated with an entire population or with certain levels of experimental factors of interest. Other models are expressed as *random effects*, where individual experimental units are drawn at random from a population, while a model with *fixed effects* and *random effects* is called a *mixed-effects* model (Pinheiro and Bates 2000).

According to Milliken and Johnson (2009), a factor is a *random effect* if its levels consist of a random sample of levels from a population of possible levels, while a factor is a *fixed effect* if its levels are selected by a nonrandom process or if its levels consist of the entire population of possible levels.

Mixed-effects models, also called multilevel models in the social science community (education, psychology, etc.), are an extension of regression models that allow for the incorporation of random effects; they are better suited to describe relationships between a response variable and some covariates in data that are grouped according to one or more classification factors. Examples of such grouped data include longitudinal data, repeated measures data, multilevel data, and block designs. One example of grouped data are animals that belong to the same herd; for example, assume we have 10 herds with 50 animals (observations) in each. By associating to observations (animals) sharing the same level of a classification factor (herd) a common random effect, mixed-effects models parsimoniously represent the covariance structure induced by the grouping of data (Pinheiro and Bates 2000). Most of the early work on mixed models was motivated by the animal science community driven by the need to incorporate heritabilities and genetic correlations in parsimonious fashion.

Next we provide an example to illustrate how to build these types of models. Assume that five environments were chosen at random from an agroecological area of Mexico. Then in each area, three replicates of a new variety (NV) of maize were tested to measure grain yield (GY) in tons per hectare. The data collected from this experiment are shown in Fig. 1.2.

Since the only factor that changes among the observations measured in this experiment is the environment, they are arranged in a one-way classification because they are classified according to a single characteristic: the environments in which the observations were made (Pinheiro and Bates 2000). The data structure is very simple since each row represents one observation for which the environment and GY were recorded, as can be seen in Table 1.1.

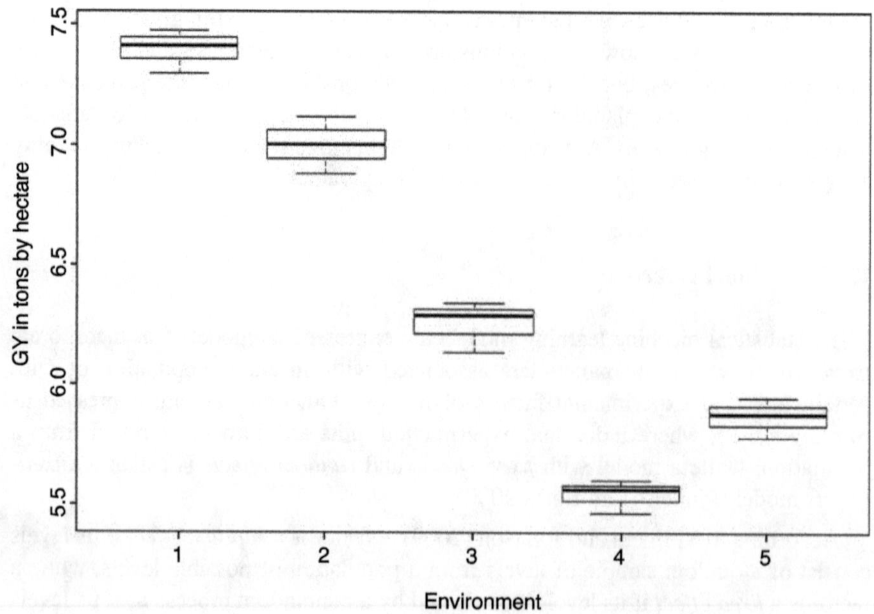

Fig. 1.2 Grain yield (GY) in tons per hectare by environment

Table 1.1 Grain yield (GY) measured in five environments (Env) with three repetitions (Rep) in each environment

Env	Rep	GY
1	1	7.476
1	2	7.298
1	3	7.414
2	1	7.117
2	2	6.878
2	3	7.004
3	1	6.136
3	2	6.340
3	3	6.288
4	1	5.600
4	2	5.564
4	3	5.466
5	1	5.780
5	2	5.948
5	3	5.881

The breeder who conducted this experiment was only interested in the average GY for a typical environment, that is, the expected GY, the variation in average GY among environments (between-environment variability), and the variation in the observed GY for a single environment (within-environment variability). Figure 1.2 shows that there is considerable variability in the mean GY for different

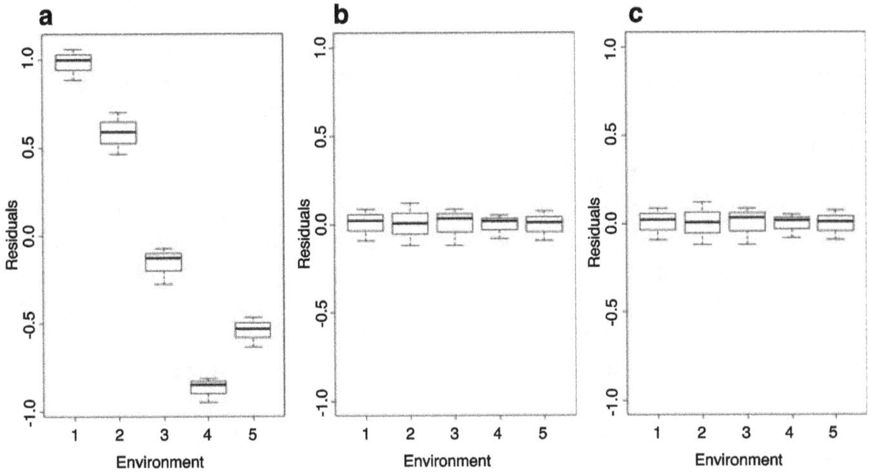

Fig. 1.3 Box plot of residuals by environments of the fit (**a**) of a single-mean model, (**b**) of a fixed-effects model with environment effect, and (**c**) of a mixed-effects model with environment effect

environments and that the within-environment variability is smaller than the between-environment variability.

Data from a one-way classification like the one given in Table 1.1 can be analyzed under both approaches with fixed effects or random effects. The decision on which approach to use depends basically on the goal of the study, since if the goal is to make inferences about the population for which these environments (levels) were drawn, then the random effect approach is the best option, but if the goal is to make inferences about the particular environments (levels) selected in this experiment, then the fixed-effects approach should be preferred.

Assuming a simple model that ignores the environment

$$GY_{ij} = \beta + e_{ij}, \qquad i = 1, .., 5, j = 1, 2, 3, \tag{1.2}$$

where GY_{ij} denotes the GY of environment i in replication j, and β is the mean GY across the population of environments sampled with $e_{ij} \sim N(0, \sigma^2)$. Using this model we estimate $\widehat{\beta} = 6.4127$ and the residual standard error is equal to $\widehat{\sigma} = 0.7197$. By fitting this model and observing the boxplot of the residuals (Fig. 1.3a) versus the environments, we can see that these residuals are very large and different for each environment, which can be attributed to the fact that this model ignores the environmental effect and was only fitted as a single-mean model, which implies that the environmental effects are included in the residuals. For this reason, we then incorporated the environmental effect in the model as a separate effect. This fixed-effects model is equal to a one-way classification model as

$$GY_{ij} = \beta_i + e_{ij}, \quad i = 1, .., 5, j = 1, 2, 3, \tag{1.3}$$

where β_i represents the fixed effect of environment i, while the other terms in the model are the same as those in Eq. (1.2). By fitting this model using least squares, we now have a beta coefficient for each environment equal to: $\widehat{\beta}_1 = 7.396$, $\widehat{\beta}_2 =6.999$, $\widehat{\beta}_3 = 6.255$, $\widehat{\beta}_4 = 5.543$, $\widehat{\beta}_5 = 5.869$, while the residual standard error is equal to $\widehat{\sigma} =0.095$. Figure 1.3b shows that the residuals are considerably lower and more centered around zero than in the first fitted model (single-mean model), which can be observed in the residual standard error that is 7.47 times smaller. Therefore, we have evidence that the model given in Eq. (1.3) successfully accounted for the environmental effects. Two drawbacks of the model with fixed effects given in Eq. (1.3) are that it is unable to provide an estimate of the between-environments variability and that the number of parameters in the model increases linearly with the number of environments. Fortunately, the random effects model circumvents these problems by treating the environmental effects as random variations around a population mean. Next we reparameterize model (1.3) as a random effects model. We write

$$GY_{ij} = \overline{\beta} + \left(\beta_i - \overline{\beta}\right) + e_{ij}, \tag{1.4}$$

where $\overline{\beta} = \sum_{i=1}^{5} \beta_i / 5$ represents the average grain yield for the environments in the experiment. The random effects model replaces $\overline{\beta}$ with the mean grain yield across the population of environments and replaces the deviations $\beta_i - \overline{\beta}$ with the random variables whose distribution is to be estimated. If $\beta_i - \overline{\beta}$ is not assumed random the model belongs to fixed effects. Therefore, the random effects version of the model given in Eq. (1.4) is equal to

$$GY_{ij} = \beta + b_i + e_{ij}, \tag{1.5}$$

where β is the mean grain yield across the population of environments, b_i is a random variable representing the deviation from the population mean of the grain yield for the ith environment, and e_{ij} is defined as before. To complete this statistical machine learning model, we must specify the distribution of the random variables b_i, with $i = 1, \ldots, 5$. It is common to assume that b_i is normally distributed with mean zero and variance between environments σ_b^2, that is, b_i is distributed $N\left(0, \sigma_b^2\right)$. It is also common to assume independence between the two random effects b_i and e_{ij}. Models with at least two sources of random variation are also called hierarchical models or multilevel models (Pinheiro and Bates 2000). The covariance between observations in the same environment is σ_b^2, which corresponds to a correlation of $\sigma_b^2 / \left(\sigma_b^2 + \sigma^2\right)$. The parameters of the mixed model given in Eq. (1.5) are β, σ_b^2, and σ^2, and irrespective of the number of environments in the experiment, the required number of parameters will always be three, although the random effects, b_i, behave like parameters. We will, however, require $\widehat{b}_i$ predictions of these random effects, given the observed data at hand. Note that when fitting this model (Eq. 1.5) for the environmental data, the parameter estimates were $\beta = 6.413$, $\sigma_b^2 = 0.594$, and $\widehat{\sigma} =$ 0.095. It is evident that there was no improvement in terms of fitting since the plot of residuals versus environment looks the same as the last fitted model (Fig. 1.3c) and the estimated residual standard error was the same as that under the fixed-effects model, which includes the effects of environment in the predictor, but as mentioned above, many times requires considerably fewer parameter estimates than a fixed-effects model.

1.4 Matrix Algebra Review

In this section, we provide the basic elements of linear algebra that are key to understanding the machinery behind the process of building statistical machine learning algorithms.

A ***matrix*** is a rectangular arrangement of numbers whose elements can be identified by the row and column in which they are located. For example, matrix $\boldsymbol{E}$, consisting of three rows and five columns, can be represented as follows:

$$\boldsymbol{E}=\begin{bmatrix} E_{11} & E_{12} & E_{13} & E_{14} & E_{15} \\ E_{21} & E_{22} & E_{23} & E_{24} & E_{25} \\ E_{31} & E_{32} & E_{33} & E_{34} & E_{35} \end{bmatrix}$$

For example, by replacing the matrix with numbers, we have

$$\boldsymbol{E}=\begin{bmatrix} 7 & 9 & 4 & 3 & 6 \\ 9 & 5 & 9 & 8 & 11 \\ 3 & 2 & 11 & 9 & 6 \end{bmatrix}$$

where the element E_{ij} is called the *ij*th element of the matrix; the first subscript refers to the row where the element is located and the second subscript refers to the column, for example, $E_{32}=2$. The order of an array is the number of rows and columns. Therefore, a matrix with r rows and c columns has an order of $r \times c$. Matrix $\boldsymbol{E}$ has an order of 3×5 and is denoted as $\boldsymbol{E}_{3\times 5}$.

In R, the way to establish an array is through the command *matrix(...)* with parameters of this function given by *matrix (data = NA, nrow = 3, ncol = 5, byrow = FALSE*) where *data* is the data for the matrix, *nrow* the number of rows, *ncol* the number of columns, and *byrow* is the way in which you will accommodate the data in the matrix by row or column. The data entered by default are FALSE, so they will fill the matrix by columns, while if you specified TRUE, they will fill the matrix by rows.

For example, to build matrix $\boldsymbol{E}$ in R, use the following R script:

$$\boldsymbol{E}=\begin{bmatrix} 7 & 9 & 4 & 3 & 6 \\ 9 & 5 & 9 & 8 & 11 \\ 3 & 2 & 11 & 9 & 6 \end{bmatrix}$$

```
E <- matrix(data= c(7,9,4,3,6,9,5,9,8,11,3,2,11,9,6), nrow = 3, ncol = 5, byrow = TRUE)
E
     [,1] [,2] [,3] [,4] [,5]
[1,]    7    9    4    3    6
[2,]    9    5    9    8   11
[3,]    3    2   11    9    6
```

To access all the values of a row, for example, the first row of matrix ***E***, you can use:

```
E[1,]
[1] 7 9 4 3 6
```

While, to access all the values of a column, for example, the first column of matrix ***E***, you can use:

```
E[,1]
[1] 7 9 3
```

To access a specific element, for example, row 3, column 2 of matrix ***E***, you can specify:

```
E[3,2]
[1] 2
```

A matrix consisting of a single row is called a vector. For example, a vector that has five elements can be represented as

$$\boldsymbol{e} = [8\ \ 10\ \ 5\ \ 4\ \ 7]$$

Here only a subscript is needed to specify the position of an element within the vector. Therefore, the *ith* element in vector ***e*** refers to the element in the ith column. For example, $e_3 = 5$.

To create a vector in R, we use the *c* (...) command that receives the data, separated by a comma. For example, the vector named ***e*** can be created using the following command:

```
e <- c(8,10,5,4,7)
e
[1] 8 10 5 4 7
```

To access a specific index, you specify the value between brackets, for example, index 3 of vector $\boldsymbol{e}$.

```
e[3]
[1] 5
```

Next we provide definitions and examples of some common types of matrices. We start with a ***square matrix***, which is a matrix with the same number of rows and columns. A matrix $\boldsymbol{B}$ of order 3×3 is shown below.

$$\boldsymbol{B} = \begin{bmatrix} 3 & 5 & 0 \\ 5 & 1 & 5 \\ -2 & -3 & 7 \end{bmatrix}$$

The ij elements in the square matrix where i equals j are called diagonal elements. The rest of the elements are known as elements that are outside the diagonal, so in this example, the elements of the diagonal of matrix $\boldsymbol{B}$ are 3, 1, and 7.

To create this type of matrix in R, simply use the *matrix* (...) command and specify the dimensions in the *ncol* and *nrow* parameters, as in the following command:

```
B <- matrix(data = c(3,5,-2,5,1,-3,0,5,7), nrow = 3, ncol = 3)
B
     [,1] [,2] [,3]
[1,]    3    5    0
[2,]    5    1    5
[3,]   -2   -3    7
```

Next we define a ***diagonal matrix*** as a square matrix that has zeros in all the elements that are outside the diagonal. For example, by extracting the diagonal elements of the above matrix ($\boldsymbol{B}$), we can form the following diagonal matrix:

$$\boldsymbol{D} = \begin{bmatrix} 3 & 0 & 0 \\ 0 & 1 & 0 \\ 0 & 0 & 7 \end{bmatrix}$$

Another special matrix is when the elements of the diagonal are 1; it is called an ***identity matrix***, and is usually denoted as $\boldsymbol{I}_r$, and r denotes the order of the matrix.

In R there is the command *diag (x = 1, ncol, nrow)* that works to create a diagonal matrix, but if you want to extract the diagonal of the $\boldsymbol{B}$ matrix, it can be extracted in the following way:

```
diag(B)
[1] 3 1 7
```

If we want to create an identity matrix, it is enough to specify the size of the matrix. For example, to create an identity matrix of order 5 × 5, the command to use would be the following:

```
I <- diag(5)
I
     [,1] [,2] [,3] [,4] [,5]
[1,]  1  0  0  0  0
[2,]  0  1  0  0  0
[3,]  0  0  1  0  0
[4,]  0  0  0  1  0
[5,]  0  0  0  0  1
```

A square matrix with all the elements above the diagonal equal to 0 is known as a *lower triangular* matrix; when all the lower elements of the diagonal are 0, it is known as an *upper triangular* matrix. In the example given below, matrix F illustrates a lower triangular matrix and matrix G illustrates an upper triangular matrix.

$$F = \begin{bmatrix} 8 & 0 & 0 \\ 10 & 7 & 0 \\ 4 & 3 & 12 \end{bmatrix}; \qquad G = \begin{bmatrix} 8 & 10 & 5 \\ 0 & 6 & 10 \\ 0 & 0 & 11 \end{bmatrix}$$

The lower triangular matrix can be extracted using the following commands:

```
F <- matrix(data = c(8,10,4,0,7,3,0,0,12), nrow = 3, ncol = 3)
F=F[upper.tri(F)] <- 0
F
     [,1] [,2] [,3]
[1,]   8   0   0
[2,]  10   7   0
[3,]   4   3  12
```

The upper triangular matrix can be extracted using the following commands:

```
G <- matrix(data = c(8,10,5,0,6,10,0,0,11), nrow = 3, ncol = 3, ,byrow=T)
G[lower.tri(G)] <- 0
G
     [,1] [,2] [,3]
[1,]    8   10    5
[2,]    0    6   10
[3,]    0    0   11
```

Next we will illustrate some basic matrix operations. We start by illustrating the ***transpose*** of matrix $\boldsymbol{E}$, commonly indicated as $\boldsymbol{E}'$ or $\boldsymbol{E}^T$; in this type of matrix, the elements *ji* are the *ij* elements of the original matrix. That is, $E'_{ji} = E_{ij}$, where the columns of $\boldsymbol{E}'$ are the rows of $\boldsymbol{E}$, and the rows of $\boldsymbol{E}'$ are the columns of $\boldsymbol{E}$. Below we provide matrix $\boldsymbol{H}$ and its transpose $\boldsymbol{H}'$.

$$\boldsymbol{H} = \begin{bmatrix} 4 & 6 \\ 6 & 2 \\ 0 & -1 \end{bmatrix}; \qquad \boldsymbol{H}' = \begin{bmatrix} 4 & 6 & 0 \\ 6 & 2 & -1 \end{bmatrix}$$

To obtain the transpose of a matrix in $\boldsymbol{H}$, the command *t (...)* is used:

```
H <- matrix(data = c(4,6,0,6,2,-1), nrow = 3,  2)
H
     [,1] [,2]
[1,]    4    6
[2,]    6    2
[3,]    0   -1
```

```
t(H)
[1,]   4   6   0
[2,]   6   2  -1
```

Two matrices can be added or subtracted only if they have the same number of rows and columns. To demonstrate the adding process, we use the following matrices:

$$\boldsymbol{J} = \begin{bmatrix} 15 & 15 \\ 25 & 35 \end{bmatrix}; \quad \boldsymbol{L} = \begin{bmatrix} 55 & 65 \\ 75 & 85 \end{bmatrix}$$

We form matrix $\boldsymbol{M}$ as the sum of matrices $\boldsymbol{J}$ and $\boldsymbol{L}$, so that $\boldsymbol{M}_{ij} = \boldsymbol{J}_{ij} + \boldsymbol{L}_{ij}$, and their sum is the following:

$$\boldsymbol{M} = \begin{bmatrix} 15+55 & 15+65 \\ 25+75 & 35+85 \end{bmatrix} = \begin{bmatrix} 70 & 80 \\ 100 & 120 \end{bmatrix}$$

Matrix $\boldsymbol{N}$ is reached by subtracting matrices $\boldsymbol{J}$ and $\boldsymbol{L}$, so $\boldsymbol{N}_{ij} = \boldsymbol{J}_{ij} - \boldsymbol{L}_{ij}$, and the subtraction is the following:

$$\boldsymbol{N} = \begin{bmatrix} 15-55 & 15-65 \\ 25-75 & 35-85 \end{bmatrix} = \begin{bmatrix} -40 & -50 \\ -50 & -50 \end{bmatrix}$$

To do the addition and subtraction of matrices in R, it is enough to use the addition or subtraction operator and fulfill the requirement that both matrices have the same dimensions. Next we reproduce the two previous addition and subtraction examples using the commands in R:

```
J <- matrix(data= c(15,15,25,35), ncol = 2, byrow = TRUE)
L <- matrix(data= c(55,65,75,85), ncol = 2, byrow = TRUE)

M<- J+L
M
     [,1] [,2]
[1,]  70  80
[2,] 100 120
```

While the subtraction of matrices is:

```
N<- J-L
N
     [,1] [,2]
[1,] -40 -50
[2,] -50 -50
```

Two matrices can be multiplied only if the number of columns in the first matrix equals the number of rows in the second. The resulting matrix will be equal to the number of rows in the first matrix and the number of columns in the second. Since $\boldsymbol{O} = \boldsymbol{PQ}$, then

$$O = \sum_{j=1}^{m} \sum_{i=1}^{n} \sum_{k=1}^{z} P_{ik} Q_{kj},$$

where m is the number of columns in matrix $\boldsymbol{Q}$, n the number of rows in matrix $\boldsymbol{P}$, and z the number of rows in $\boldsymbol{Q}$ and number of columns in $\boldsymbol{P}$. To demonstrate the above, we have

$$\boldsymbol{P} = \begin{bmatrix} 6 & 8 & 10 \\ 5 & 6 & 8 \\ 4 & 6 & 9 \end{bmatrix}; \boldsymbol{Q} = \begin{bmatrix} 3 & 5 \\ 2 & 2 \\ 9 & 8 \end{bmatrix}$$

Then $\boldsymbol{S}$ is obtained as

$$\begin{aligned} S_{11} &= 6 \times 3 + 8 \times 2 + 10 \times 9 = 124 \\ S_{21} &= 5 \times 3 + 6 \times 2 + 8 \times 9 = 99 \\ S_{31} &= 4 \times 3 + 6 \times 2 + 9 \times 9 = 105 \\ S_{12} &= 6 \times 5 + 8 \times 2 + 10 \times 8 = 126 \\ S_{22} &= 5 \times 5 + 6 \times 2 + 8 \times 8 = 101 \\ S_{32} &= 4 \times 5 + 6 \times 2 + 9 \times 8 = 104 \end{aligned}$$

Therefore,

$$\boldsymbol{S} = \begin{bmatrix} 124 & 126 \\ 99 & 101 \\ 105 & 104 \end{bmatrix}$$

Note that $\boldsymbol{S}$ is of order 3×2, where 3 represents the number of rows in $\boldsymbol{P}$ and 2 equals the number of columns in $\boldsymbol{Q}$.

In order to multiply matrices in R, it is necessary to use the operator %*% between the two matrices, in addition to meeting the requirements mentioned above.

```
P <- matrix(data=c(6,5,4,8,6,6,10,8,9), ncol=3)
Q <- matrix(data=c(3,2,9,5,2,8), ncol=2)

S <- P%*%Q
S
     [,1] [,2]
[1,] 124 126
[2,]  99 101
[3,] 105 104
```

The inverse of a matrix usually is denoted as $\boldsymbol{R}^{-1}$, and when it is multiplied by the original matrix, it results in an identity matrix, that is, $\boldsymbol{R}^{-1}\boldsymbol{R} = \boldsymbol{I}$, where $\boldsymbol{I}$ is the identity matrix. Only square matrices are invertible.

If it is a diagonal matrix, its inverse can be calculated simply in the following way:

$$\boldsymbol{R} = \begin{bmatrix} 7 & 0 & 0 \\ 0 & 8 & 0 \\ 0 & 0 & 6 \end{bmatrix}$$

Then

$$\boldsymbol{R}^{-1} = \begin{bmatrix} \frac{1}{7} & 0 & 0 \\ 0 & \frac{1}{8} & 0 \\ 0 & 0 & \frac{1}{6} \end{bmatrix}$$

For a square matrix of 2×2, its inverse can be calculated by finding the determinant, which is the difference between the product of the two elements of the diagonals and the product of the two elements outside the diagonal ($R_{11}R_{22} - R_{12}R_{21}$). Then the position of the elements of the diagonals is reversed by multiplying the elements outside the diagonal by −1 and dividing all the elements by the determinant.

$$\boldsymbol{R} = \begin{bmatrix} 4 & 2 \\ 1 & 6 \end{bmatrix}$$

Then

$$\boldsymbol{R}^{-1} = \frac{1}{(4 \times 6) - (1 \times 2)} \begin{bmatrix} 6 & -2 \\ -1 & 4 \end{bmatrix} = \begin{bmatrix} 0.2727 & -0.0909 \\ -0.0455 & 0.1818 \end{bmatrix}$$

To obtain the inverse in ***R***, the solve(...) command is the function used to perform this process, as shown in the following example, where the inverse of the matrix ***R*** is obtained.

```
R <- matrix(data = c(4, 1, 2, 6), ncol = 2)
R
     [,1] [,2]
[1,]    4    2
[2,]    1    6
```

```
solve(R)
        [,1]     [,2]
[1,] 0.2727 -0.0909
[2,] -0.0455 0.1818
```

1.5 Statistical Data Types

1.5.1 Data Types

To use statistical learning methods correctly, it is very important to understand the classification of the types of data that exist. This is of paramount importance because data are the input to all statistical machine learning methods and because the data type *determines* the appropriate and valid analysis to be implemented; in addition, each statistical machine learning method is specific to a certain type of data. In general, data are most commonly classified as quantitative (numerical) or qualitative (categorical) (Fig. 1.4).

By quantitative (numerical) data, we understand that the result of the observation or the result of a measurement is a number. They are classified as

(a) Discrete. The variable can only have point values and no values in between, that is, the variable can only have a certain set of possible values and represent items that can be counted because they only have isolated numerical values. Examples: number of household members, number of surgical interventions, number of reported cases of a certain pathology, number of accidents per month, etc. Examples in the context of plant breeding are panicle number per plant, seed number per panicle, weed count per plot, number of infected spikelets per spike, etc. Also, discrete values are called as count responses and those models based on Poisson and negative binomial distribution are appropriate for this type of responses.

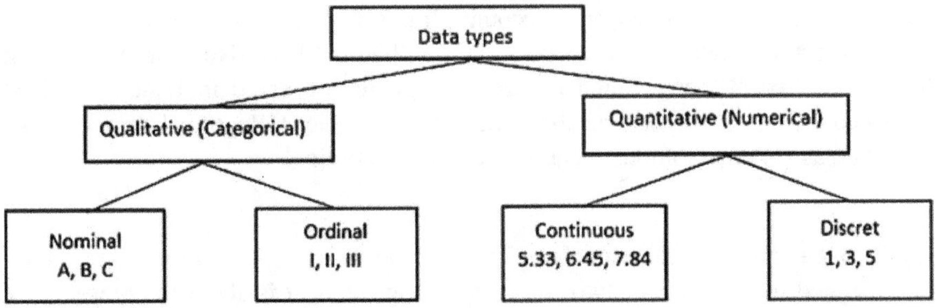

Fig. 1.4 Types of data

(b) Continuous. They are usually the result of a measurement that is expressed in particular units, and values are measured based on a zero point and are treated as real numbers. There are many types of mathematical operations that can be performed on this type of data. The measurements can theoretically have an infinite set of possible values within a range and they do not need transformation. In practice, the possible values of the variable are limited by the accuracy of the measurement method or by the recording mode. Examples: plant height, age, weight, grain yield, pH, blood cholesterol level, etc. The distinction between discrete and continuous data is important for deciding which statistical learning method to use for the analysis, since there are methods that assume that the data are continuous. Consider, for example, the age variable. Age is continuous, but if it is recorded in years, it turns out to be discrete. In studies with adults, in which the age ranges from 20 to 70 years, for example, there are no problems in treating age as continuous, since the number of possible values is large. But in the case of preschool children, if the age is recorded in years, it should be treated as discrete, while if it is recorded in months, it can be treated as continuous.

Similarly, the variable number of beats per minute is a discrete variable, but it is treated as continuous due to the large number of possible values. Numerical data (discrete or continuous) can be transformed into categorical and be treated as such. Although this is correct, it is not necessarily efficient because information is lost during the categorization process. It is always preferable to record the numerical value of the measurement, since this makes it possible to (a) analyze the variable as numerical because statistical analysis is simpler and more powerful and (b) form new categories using different criteria. Only in special cases it is preferable to record numerical data as categorical, for example, when the measurement is known to be imprecise (number of cigarettes per day, number of cups of coffee per week).

Categorical variables result from registering the presence of an attribute. The categories of a qualitative variable must be clearly defined during the design stage of the research and must be mutually exclusive and exhaustive. This means that each observation unit must be classified unambiguously in one, and only one, of the possible categories and that there is a category to classify each individual. In this sense, it is important to take into account all possibilities when constructing categorical variables, including a category such as "Do not know/No answer," or "Not Registered," or "Other," which ensures that all the observed individuals will be classified based on the criteria that define the variable. Categorical data are also classified as (a) dichotomous, (b) nominal, and (c) ordinal.

(a) Two categories (dichotomous). The individual or observation unit can be assigned to only one of two categories. In general, it is about the presence or absence of the attribute and it is advantageous to assign code 0 to the absence and 1 to the presence.

(b) Examples: (1) resistance—no resistance, (2) disease—no disease, (3) tall—not tall, and (4) red color—no red color. It should be noted that examples 1 and 2 definitely cover all categories, while 3 and 4 are simplifications of more complex categories. In 3 and 4 it was necessary to establish a cutoff criterion to assemble a categorical variable from a numerical variable.

(c) More than two categories. When there are more than two categories, data can be nominal or ordinal. In nominal categories, there is no obvious order between the categories. These types of data values are distinct symbols, and these values serve as labels. The term "nominal" comes from the latin word for "name." Nominal attributes or labels have no relation to one another, nor is any order implied (Patterson and Gibson 2017). Some examples are religion: Catholicism, Islam, Judaism, etc.; race type: African, American, European, Asian, other; type of species; location; plant color; etc. In ordinal data, there is an obvious order between categories. Ordinal values have rank, giving us a notion of order but no concept of distance between the values. We can compare ordinal values with one another, but mathematical operations don't make sense in the context of these values (Patterson and Gibson 2017). Some examples are

1. Drought resistance: no resistance/low resistance/medium resistance/high resistance/total resistance
2. Disease severity: absent/mild/moderate/severe
3. Temperature of a process: hot/mild/cool
4. Social class: lower/middle/upper

Even when ordinal data can be coded as numbers as in the case of stages of drought resistance from 1 to 5 (1 = no resistance, 2 = low resistance, 3 = medium resistance, 4 = high resistance, 5 = total resistance), we cannot say that a plant in stage 4 has a drought resistance twice as strong as the resistance of a plant in stage 2, nor that the difference between stages 1 and 2 is the same as between stages 3 and 4. In contrast, when considering the age of a person, 40 years is twice 20 and a difference of 1 year is the same across the entire range of values. Therefore, we need to be aware that in ordinal data the difference between categories does not make sense.

Ordinal traits are very common in plant breeding programs for measuring disease incidence and severity and for sensory evaluation, such as the perceived quality of a product (e.g., taste, smell, color, decay) and plant development (e.g., developmental stages, maturity). These types of data are often partially subjective since the scale indicates only relative order and no absolute amounts; therefore, the intervals between successive categories may not be the same (Simko and Piepho 2011).

For this reason, we must be careful when dealing with qualitative variables, especially when they have been coded numerically, since they cannot be analyzed as numbers but must be analyzed as categories. It is incorrect to present, for example, the average stage of drought resistance in a group of plants.

In practice, scales are used to define degrees of a symptom or a disease, such as 1, 2, 3, 4, and 5. For this reason, it is important to operationally define this type of variables and study their reliability in order to ensure that two observers placed in front of the same plant will classify it in the same category.

Table 1.2 Examples of multivariate data

Units	Variables	Types of data
Plant	Several measurements of plant height on a single plant in time	All continuous
Animal	Measurement of three animal traits (average daily weight gain, muscularity, and calving)	Mixture of continuous and ordinal
Students	Grades in mathematics, physics, chemistry, biology	All continuous
People	Income, type of residence, gender, educational level, occupation	Mixture of nominal, ordinal, and continuous
Wheat plant	Measurement of four traits: grain yield, panicle number per plant, drought resistance, and type of wheat (common or durum)	Mixture of continuous, discrete, ordinal, and nominal
Country	Several measurements of school performance using the programme for international student assessment (PISA) test	Exam scores continuous in mathematics, reading and science

1.5.2 Multivariate Data Types

Practitioners and researchers in all applied disciplines often measure several variables in each observation, subject, unit, or experimental unit. That is, all variables are simultaneously measured in the same observation. Multivariate data are very common in all disciplines due to the need and facility for data collection in most fields. These variables can consist of only one type of data (for example, plant height measured using a continuous scale on each plant 12 times every 15 days) or a mixture of data types, for example, measuring, on each plant, four different traits: grain yield (on a continuous scale), disease resistance (on an ordinal scale), flower color (nominal scale), and days to flowering (discrete or count). Table 1.2 provides other examples of multivariate data measured using only one scale or a mixture of scales.

It is important to point out that here all measurements are done simultaneously in each observation. For this reason, they are classified as multivariate type of data and include data that will be used as dependent variables or independent variables in the process of training the statistical machine learning algorithms that will be studied here.

1.6 Types of Learning

The three most common ways of learning in statistical machine learning are (a) supervised learning, (b) unsupervised learning, and (c) semi-supervised learning. The three methods are explained below.

1.6.1 Definition and Examples of Supervised Learning

Supervised learning can be defined as the process of learning a function that maps an input to an output based on teaching the statistical machine learning method with input–output pairs. The training data consist of pairs of objects (usually vectors): one component of the pair is the input data (predictors = explanatory variable = input) and the other, the desired results (response variable = dependent variable = output). The output of the function can be a numerical value (as in regression problems) or a class label (as in multinomial regression). The goal of supervised learning is to learn a function that, given a sample of data and desired outputs, best approximates the relationship between input and output observable in the data. This function should be capable of predicting the value corresponding to any valid input object after having seen a series of examples of training data. Under optimal conditions, the algorithm correctly determines the class labels for unseen instances. This implies a learning algorithm that is able to generalize from the training data to unseen situations in a "reasonable" way.

Suppose you're teaching your child to distinguish between corn and tomato (Fig. 1.5). First you show him (her) a picture of an ear of corn and a picture of a tomato. In the learning process, your child must keep in mind that if the color is yellow and the shape is not round, then it is probably an ear of corn, but if the color is red and the shape is round, then it is probably a tomato. This is how your child learns. Then you can show a third picture and ask your child to classify the vegetable as either ear of corn or tomato. When you show the third picture, he (she) will very likely identify if the vegetable is ear of corn or tomato, due to the fact that we have already labeled the two pictures into categories, so your child knows what is an ear of

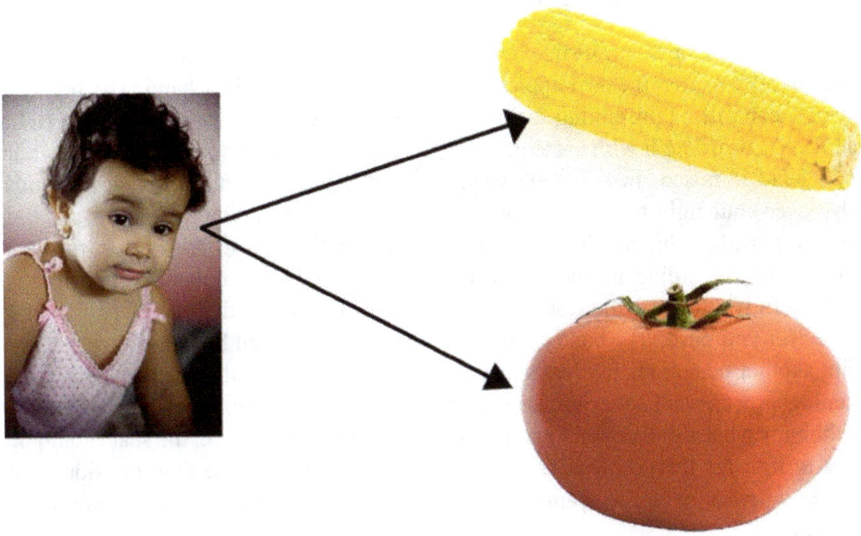

Fig. 1.5 Supervised learning process for teaching a child to distinguish tomato from corn

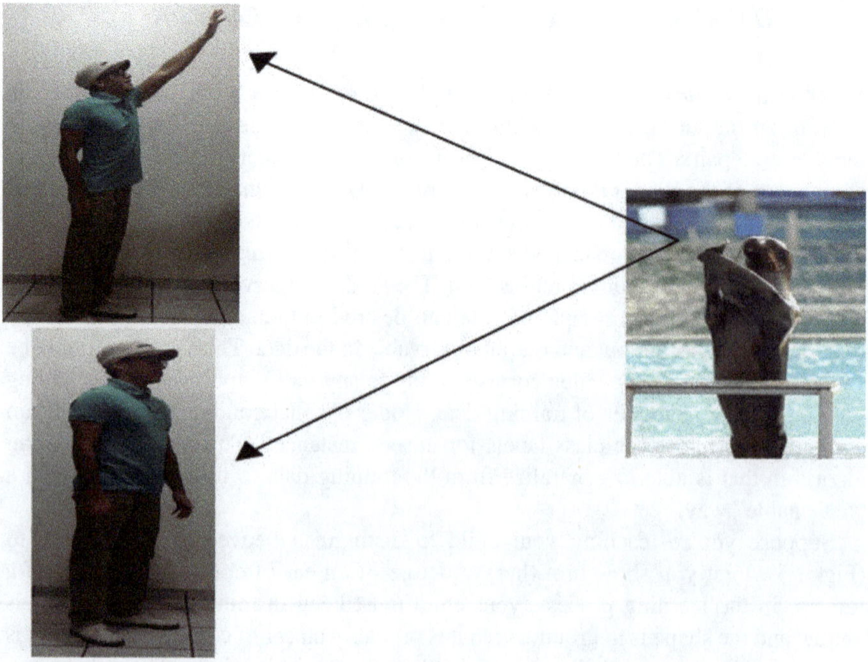

Fig. 1.6 Supervised learning process for teaching a seal to applaud

corn and what is a tomato. This example illustrates how supervised learning works using ground truth data that consist of having prior knowledge of what the output values of our samples should be.

To give another example, imagine that you're training a seal to applaud (Fig. 1.6). The goal is to make the seal applaud when you raise your right hand. The training process consists of presenting the seal with enough examples by raising your right hand and rewarding it with some great candy whenever it applauds when it sees your right hand is raised. In the same way, the seal may be "punished" if it applauds whenever your right hand is not raised, by doing something unpleasant for the seal but not harmful. Supervision involves stimulating the seal to respond to positive samples by rewarding it, and not to respond to negative samples by "punishing" it. Hopefully, the seal then obtains a built-in feeling (hypothesis) for applauding whenever you raise your hand right. The process is evaluated by presenting the seal with another person raising his/her right hand, someone who did not take part in the training process and who is unknown to the seal. However, based on its built-in feeling for what a person with his/her raised right hand looks like, the seal should be able to transfer this knowledge to the present person. It must then consider and decide whether or not it wants to signal the presence of a right hand raised by applauding.

Next we provide some real examples. In the first example, a scientist has thousands of molecules and information about which ones are drugs and he trains

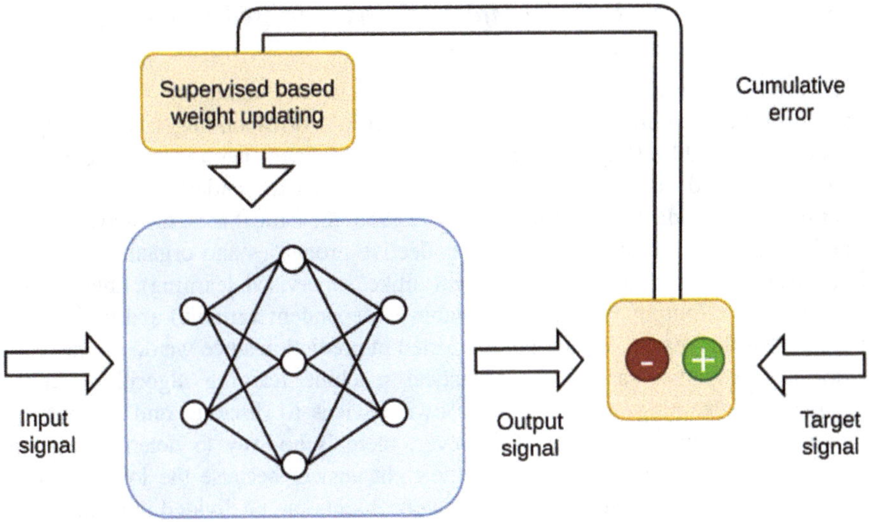

Fig. 1.7 Diagram illustrating a supervised learning process where there are inputs (X) and response variables (y) for each observation

a statistical machine learning model to determine whether a new molecule is also a drug. In the context of plant breeding, a scientist collects hundreds of markers (genetic information) and thousands of images of hundreds of plants. For this sample of plants, he also measures their phenotype (grain yield) because he wants to implement a statistical machine learning algorithm to estimate (predict) the grain yield of new plants not used in the training process. Another example is in environmental science, where scientists use historical data (as when it's sunny and the temperature is higher, or when it's cloudy and the humidity is higher, etc.) to train a statistical machine learning model to predict the weather for a given future time.

In a more mathematical way, under supervised learning, we usually have access to a set of p predictor (input) variables $X_1, X_2, \ldots, X_p$ measured in n observations, and a response variable (output) Y also measured in those same n observations (Fig. 1.7). The goal is to predict Y using a function of $X_1, X_2, \ldots, X_p$, that is, we use an algorithm to learn the mapping function from the input to the output $Y = f(X_1, X_2, \ldots, X_p)$, and we expect to estimate the mapping function so well that when we have new input data, we can predict the output variables for those data. The term supervised learning was coined because the learning process of any statistical machine learning method from the training dataset can be thought of as a teacher supervising the learning process. We know the correct outputs (response variables), the algorithm iteratively makes predictions on the training data and is corrected by the teacher. Learning stops when the statistical machine learning algorithm achieves an acceptable level of performance.

1.6.2 Definitions and Examples of Unsupervised Learning

Unsupervised learning is when you only have input (predictors = independent variables) data (*X*) and no previous knowledge of corresponding labeled outputs or response variables (Fig. 1.8). So its goal is to deduce the natural structure present within a set of data points. In other words, to extract the underlying structure or distribution in the data in order to learn more about the data, that is, the network uses training patterns to discover emerging collective properties and organizes the data into clusters. In unsupervised learning (unlike supervised learning), there is no correct answer (output = response variable = dependent variable) and there is no teacher. For this reason, we are not interested in prediction since we do not have an associated response variable *Y*. Statistical machine learning algorithms under unsupervised learning are left to their own devices to discover and present the interesting structure in the data. However, there is no way to determine if our work is correct since we don't know the right answer because the job was done without supervision. Unsupervised learning problems can be divided into clustering and association problems.

Clustering: A clustering problem is when you want to discover the inherent groupings in the data, such as grouping maize hybrids by their genetic architecture. Another example is grouping people according to their consumption behaviors. But in both cases we cannot check if the classifications are correct since we don't know the true grouping of each individual.

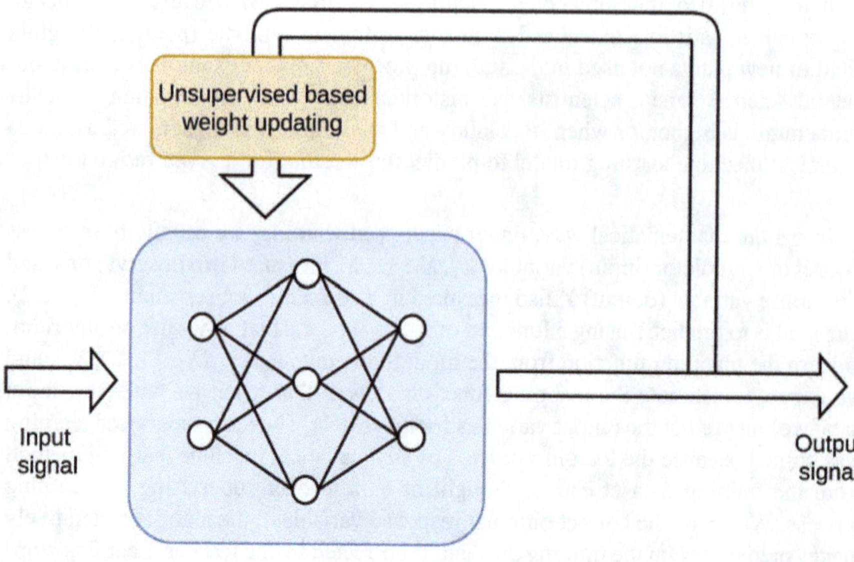

Fig. 1.8 Diagram illustrating an unsupervised learning process where there are only inputs (X), but no response variables (y) for each observation

Association: An association rule learning problem is when you want to discover rules that describe large portions of your data, such as people who buy X also tend to buy Y.

Some popular examples of unsupervised learning algorithms are

(a) Principal component analysis
(b) Multidimensional scaling for grouping
(c) A k-means algorithm for clustering problems
(d) An a priori algorithm for association rule learning problems

1.6.3 Definition and Examples of Semi-Supervised Learning

Semi-supervised learning problems are those that have a large amount of input data (X) available but only some of the data are labeled (Y). For this reason, these problems are positioned between supervised and unsupervised learning. A good example is plant species classification using thousands of images where only some of the images are labeled (e.g., species 1, species 2, species 3, etc.) and the majority are unlabeled. Another example is the classification of exoplanets (exoplanets are planets that are outside our solar system) also using thousands of photos where only a small fraction of the photos is labeled (four types of exoplanets). Many real-world problems in the context of statistical machine learning belong to this type of learning process. This is because it is more expensive and time-consuming to use labeled data than unlabeled data since many times this requires having access to domain experts, whereas it is cheap and easy to collect and store unlabeled data.

References

Adler I (1912) Primary malignant growths of the lungs and bronchi: a pathological and clinical study. Longmans, Green and Co, New York and London, p 325

Bernardo R (2016) Bandwagons I, too, have known. Theor Appl Genet. https://doi.org/10.1007/s00122-016-2772-5

Box GEP (1976) Science and statistics (PDF). J Am Stat Assoc 71:791–799. https://doi.org/10.1080/01621459.1976.10480949

Breiman L (2001) Statistical modeling: the two cultures. Stat Sci 16:199–215

Crossa J, Pérez-Rodríguez P, Cuevas J, Montesinos-López OA, Jarquín D, de Los Campos G, Burgueño J, González-Camacho JM, Pérez-Elizalde S, Beyene Y, Dreisigacker S, Singh R, Zhang X, Gowda M, Roorkiwal M, Rutkoski J, Varshney RK (2017) Genomic selection in plant breeding: methods, models, and perspectives. Trends Plant Sci 22(11):961–975

Dean J (2018) Big data, data mining, and machine learning. Value creation for business leaders and practitioners. John Wiley & Sons, Inc., Hoboken

FAO (2011) The state of the World's land and water resources for food and agriculture: managing Systems at Risk. Food and agriculture Organization of the United Nations. FAO, Rome

Fischer T, Byerlee D, Edmeades G (2014) Crop yields and global food security. ACIAR, Canberra

James G, Witten D, Hastie T, Tibshirani R (2013) An introduction to statistical learning: with applications in R. Springer, New York

McKinsey Global Institute (2016) The age of analytics: competing in a data-driven world. https://www.mckinsey.com/~/media/mckinsey/business%20functions/mckinsey%20analytics/our%20insights/the%20age%20of%20analytics%20competing%20in%20a%20data%20driven%20world/mgi-the-age-of-analytics-executive-summary.ashx

Meuwissen THE, Hayes BJ, Goddard ME (2001) Prediction of total genetic value using genome-wide dense marker maps. Genetics 157:1819–1829

Milliken GA, Johnson DE (2009) Analysis messy of data, volume 1 designed experiments. CRC Press Taylor & Francis Group, Boca Raton, London, New York

Montesinos-López A, Martín-Vallejo J, Crossa J, Gianola D, Hernández-Suárez CM, Montesinos-López OA, Juliana P, Singh R (2019) A benchmarking between deep learning, support vector machine and Bayesian threshold best linear unbiased prediction for predicting ordinal traits in plant breeding. G3 9(2):601–618

Oury F-X, Godin C, Mailliard A, Chassin A, Gardet O, Giraud A, Heumez E, Morlais J-Y, Rolland B, Rousset M, Trottet M, Charmet G (2012) A study of genetic progress due to selection reveals a negative effect of climate change on bread wheat yield in France. Eur J Agron 40:28–38

Patterson J, Gibson A (2017) Deep learning: a Practitioner's approach. O'Reilly Media, Beijing

Pinheiro JC, Bates DM (2000) Mixed-effects models in S and S-PLUS. Springer Verlag, New York

Proctor RN (2012) The history of the discovery of the cigarette-lung cancer link: evidentiary traditions, corporate denial, global toll. Tob Control 21(2):87–91

Samuel AL (1959) Some studies in machine learning using the game of checkers. IBM J Res Dev 3 (3):210–229

Schuster PM (2014) The scientific life of Victor Franz (Francis) Hess (June 24, 1883–December 17, 1964). Astropart Phys 53:33–49

Sejnowski TJ (2018) The deep learning revolution. The MIT Press, Cambridge, MA, London

Shmueli G (2012) To explain or to predict? Stat Sci 25(3):289–310. https://doi.org/10.1214/10-STS330

Simko I, Piepho H-P (2011) Combining phenotypic data from ordinal rating scales in multiple plant experiments. Trends Plant Sci 16:235–237

Stroup W (2012) Generalized linear mixed models: modern concepts, methods and applications. CRC Press, Boca Raton

Wang X, Xua Y, Hu Z, Hu C (2018) Genomic selection methods for crop improvement: current status and prospects. Crop J 6(4):330–340

Wolpert DH (1996) The lack of a priory distinction between learning algorithms. Neural Comput 8 (7):1341–1390

2

Preparing Data through Preprocessing Tools

2.1 Fixed or Random Effects

As mentioned in Chap. 1, fixed effect models in general (to design experiments, regression models, genomic prediction models, etc.) are recommended when the levels under study (collected by the scientist) are the unique levels of interest in the study, and the levels or quantities observed in the explanatory variables are treated as if they were nonrandom. For these reasons, a fixed factor is defined as a categorical or classification variable, chosen to represent specific conditions, for which the researcher has included all levels (or conditions) that are of interest for the study in the model. This means that fixed effects are unknown constant parameters associated with continuous covariates or levels of categorical factors in any fixed or mixed effects (fixed + random effects). The estimation of these fixed parameters in fixed effects models or mixed effects models is generally of intrinsic interest, since they indicate the relationships of the covariates with the response variable. Fixed effects can be associated with continuous covariates such as the weight of an animal in kilograms, maize yield in tons per hectare, qualification of a reference test or socioeconomic level, which will carry a continuous range of values, or they can be associated with factors such as gender, hybrid, or group treatment, which are categorical. This implies that fixed effects are the best option when performing an inference for the whole target population.

A random factor is a classification variable with levels that can be randomly sampled from a population with different levels of study, such as classrooms, regions, cattle herds, or clinics that are randomly sampled from a population. All possible levels of the random factor are not present in the data set, yet it is the intention of the researcher to make an inference about the entire population of levels from the selected sample of these factor levels. Random factors are included in an analysis in order for the modification in the dependent variable through the levels of the random factors to be evaluated and the results of the data analysis generalized to all levels of the population random factor. This means that random effects are represented by random variables (not observed) which, we generally assume, have a particular distribution, the normal distribution being the most common. Due to the above, random effects are suggested when we want to perform an inference for all levels of the target population.

2.2 BLUEs and BLUPs

This section presents the concepts and terminologies of BLUE and BLUP. Since these two concepts are related to a mixed model, we present the following linear mixed model as

$$\boldsymbol{Y} = \boldsymbol{X\beta} + \boldsymbol{Zu} + \boldsymbol{\varepsilon}, \tag{2.1}$$

where $\boldsymbol{Y}$ is the vector of response variables of order $n \times 1$, $\boldsymbol{X}$ is the design matrix of fixed effects of order $n \times p$, $\boldsymbol{\beta}$ is the vector of order $p \times 1$ of beta coefficients, $\boldsymbol{Z}$ is the design matrix of random effects of order $n \times q$, $\boldsymbol{u}$ is the vector of random effects distributed as $N(\boldsymbol{0}, \boldsymbol{\Sigma})$, where $\boldsymbol{\Sigma}$ is a variance–covariance matrix of random effects of dimension $q \times q$, and $\boldsymbol{\varepsilon}$ is a vector of residuals distributed as $N(\boldsymbol{0}, \boldsymbol{R})$, where $\boldsymbol{R}$ is a variance–covariance matrix of residual effects of dimension $n \times n$. The unconditional mean of $\boldsymbol{Y}$ is equal to $E(\boldsymbol{Y}) = \boldsymbol{X\beta}$, while the conditional mean of $\boldsymbol{Y}$, given the random effects, is equal to $E(\boldsymbol{Y}|\boldsymbol{u}) = \boldsymbol{X\beta} + \boldsymbol{Zu}$. A solution to jointly "estimate" parameters $\boldsymbol{\beta}$ and $\boldsymbol{u}$ was proposed by Henderson (1950, 1963, 1973, 1975, 1984), which consists in solving the mixed model equation (MME)

$$\begin{pmatrix} \boldsymbol{X}^{\mathrm{T}}\boldsymbol{R}^{-1}\boldsymbol{X} & \boldsymbol{X}^{\mathrm{T}}\boldsymbol{R}^{-1}\boldsymbol{Z} \\ \boldsymbol{Z}^{\mathrm{T}}\boldsymbol{R}^{-1}\boldsymbol{X} & \boldsymbol{Z}^{\mathrm{T}}\boldsymbol{R}^{-1}\boldsymbol{Z} + \boldsymbol{\Sigma}^{-1} \end{pmatrix} \begin{pmatrix} \widehat{\boldsymbol{\beta}} \\ \widehat{\boldsymbol{u}} \end{pmatrix} = \begin{pmatrix} \boldsymbol{X}^{\mathrm{T}}\boldsymbol{R}^{-1}\boldsymbol{y} \\ \boldsymbol{Z}^{\mathrm{T}}\boldsymbol{R}^{-1}\boldsymbol{y} \end{pmatrix} \tag{2.2}$$

The solution obtained for $\boldsymbol{\beta}$ is the BLUE and the solution obtained for $\boldsymbol{u}$ is the BLUP.

While this expression to find the estimates of $\widehat{\boldsymbol{\beta}}$ and $\widehat{\boldsymbol{u}}$ may look quite complex, when the number of observations is larger than the sum of the number of fixed effects and the number of random effects $(p + q)$, it is quite efficient since only needs to calculate the inverse of the small matrices of $\boldsymbol{R}$ and $\boldsymbol{\Sigma}$. Also, the matrix on the left that needs to be inverted to obtain the solution for $\widehat{\boldsymbol{\beta}}$ and $\widehat{\boldsymbol{u}}$ is of order $(p+q) \times (p+q)$, which in some applications is considerably less than a matrix of dimension $n \times n$ as $\boldsymbol{V} = \boldsymbol{Z\Sigma Z}^{\mathrm{T}} + \boldsymbol{R}$, which is also useful to obtain the solution of these parameters by using $\widehat{\boldsymbol{\beta}} = (\boldsymbol{X}^{\mathrm{T}}\boldsymbol{V}^{-1}\boldsymbol{X})^{-1}\boldsymbol{X}^{\mathrm{T}}\boldsymbol{V}^{-1}\boldsymbol{y}$ and $\widehat{\boldsymbol{u}} = \boldsymbol{\Sigma Z}^{\mathrm{T}}\left(\boldsymbol{y} - \boldsymbol{X}\widehat{\boldsymbol{\beta}}\right)$. Under both solutions, for $\widehat{\boldsymbol{\beta}}$ and $\widehat{\boldsymbol{u}}$ it is assumed that the covariance matrices are known, but in practice these are replaced by estimations and the results are known as empirical BLUE (EBLUE) and empirical BLUP (EBLUP).

The linear combinations of the parameters are called **estimable functions** if they can be constructed from a linear combination of unconditional means (of fixed effects only) of the observations (Littell et al. 1996). Estimable functions do not depend on random effects. Below, we provide a formal definition of an estimable function.

Definition of an estimable function $\boldsymbol{K}^{\mathrm{T}}\boldsymbol{\beta}$ is estimable if there is a matrix $\boldsymbol{T}$ such that

$$\boldsymbol{T}^{\mathrm{T}}\boldsymbol{E}(\boldsymbol{Y}) = \boldsymbol{T}^{\mathrm{T}}\boldsymbol{X}\boldsymbol{\beta} = \boldsymbol{K}^{\mathrm{T}}\boldsymbol{\beta}\,\forall\,\boldsymbol{\beta}.$$

One way of testing candidate matrices for estimability is to use the following result:

$$\boldsymbol{K}^{\mathrm{T}}\boldsymbol{\beta} \text{ is estimable if, and only if, } \boldsymbol{K}^{\mathrm{T}}\left(\boldsymbol{X}^{\mathrm{T}}\boldsymbol{X}\right)^{-}\left(\boldsymbol{X}^{\mathrm{T}}\boldsymbol{X}\right) = \boldsymbol{K}^{\mathrm{T}},$$

where $(\boldsymbol{X}^{\mathrm{T}}\boldsymbol{X})^{-}$ denotes a generalized inverse of $\boldsymbol{X}^{\mathrm{T}}\boldsymbol{X}$. Quantities such as regression coefficients, treatment means, treatment differences, contrasts, and simple effects in factorial experiments are all common examples of estimable functions and their resulting estimates are examples of BLUEs (Littell et al. 2006). BLUEs correspond to broad inference because they are valid for the whole population under study, and are also called population average inference using the terminology by Zeger et al. (1998). Table 2.1 presents predictors of some regression models and some common experimental designs and also provides some functions of the predictor that are and are not estimable functions.

Table 2.1 indicates that estimable functions used to obtain BLUEs are only linear combinations of fixed effects, and the inference focuses on the average performance throughout the target population. Although there are many possible linear combinations of fixed effects that can be of interest to estimate with BLUEs, the most important ones are treatment means expressed as $\eta_0 + \tau_i$, the difference between treatments $\tau_i - \tau_{i'}$ and simple effects. In general, the most common BLUEs can be obtained from any fitted generalized mixed model using the expression BLUE $=$ $g^{-1}\left(\boldsymbol{X}\widehat{\boldsymbol{\beta}}\right)$, where $g^{-1}(\cdot)$ is the inverse link used to fit the generalized mixed model and the inference is related to population-wide average (broad).

Estimability matters because many models are not full rank, such as analysis of variance (ANOVA) models, where estimating equation solutions for the effects themselves has no intrinsic meaning and solutions depend entirely on the generalized inverse used. Theory says that there is an infinite number of ways to construct a generalized inverse. While estimable functions are invariant of generalized inverse and therefore have an assignable meaning, that is, although the effect estimates per se do not have any legitimate interpretation, estimable functions do (Stroup 2012). Next, an example of how to obtain the BLUEs of genotypes (treatments) under a randomized complete block design is provided.

Table 2.1 Predictors and some estimable and non-estimable functions of each predictor of some common useful linear models

Model	Predictor	Estimable functions	Non-estimable functions
Complete randomized design (CRD)	$\eta_i = \eta_0 + \tau_i$	$\eta_0 + \tau_i$, $\tau_i - \tau_{i'}, i \neq i'$	τ_i, η_0
Randomized complete blocks (RCBD)	$\eta_{ij} = \eta_0 + \tau_i + b_j$, $b_j \sim N(0, \sigma_b^2)$	$\eta_0 + \tau_i$, $\tau_i - \tau_{i'}, i \neq i'$	τ_i, η_0
Regression	$\eta_i = \eta_0 + \sum_{j=1}^{p} x_{ij}\beta_j$	$\eta_0 + \sum_{j=1}^{p} x_{ij}\beta_j$, $\eta_0 + x_{ij}\beta_j$	η_0
Split plot design in a CRD	$\eta_{ijk} = \eta_0 + \alpha_i + \alpha(r)_{ik} + \beta_j + (\alpha\beta)_{ij}$; $\beta_j \sim N\left(0, \sigma_\beta^2\right)$, $(\alpha\beta)_{ij} \sim N\left(0, \sigma_{\alpha\beta}^2\right)$, $\alpha(r)_{ik} \sim N\left(0, \sigma_{\alpha(r)}^2\right)$	$\eta_0 + \alpha_i$	η_0
Split plot design in RCBD	$\eta_{ijk} = \eta_0 + \alpha_i + r_k + \alpha(r)_{ik} + \beta_j + (\alpha\beta)_{ij}$; $\beta_j \sim N\left(0, \sigma_\beta^2\right)$, $(\alpha\beta)_{ij} \sim N\left(0, \sigma_{\alpha\beta}^2\right)$, $r_k \sim N(0, \sigma_r^2)$ $\alpha(r)_{ik} \sim N\left(0, \sigma_{\alpha(r)}^2\right)$	$\eta_0 + \alpha_i$	η_0

Example 1 Grain yield of five genotypes evaluated in a randomized complete block design. The data of this experiment are shown in Table 2.2. This example is provided to illustrate the process of estimating the BLUEs of genotypes.

Next, we provide the R code that uses the lme4 library to fit a mixed model for the data given in Table 2.2 to estimate the BLUEs of genotypes:

```
Data_RCBD=read.table("Example_RCBD.csv", header =T, sep = ",")
Data_RCBD

library(lme4)
Data_RCBD$Genotype=as.factor(Data_RCBD$Genotype)
Data_RCBD$Block=as.factor(Data_RCBD$Block)

Fitted=lmer(Yield~ Genotype + (1 | Block), Data_RCBD)
Fitted
####Extracting design matrix of fixed effects (Intercept and Genotype)
X=Fitted@pp$X
X=X[!duplicated(X), ]
X
####Extracting the beta coefficients
Beta=Fitted@beta #fixef(Fitted)
####Obtaining the BLUEs of genotypes
BLUEs_Gen=X%*%Beta
BLUEs_Gen
```

Table 2.2 Grain yield (Yield) of five genotypes under a randomized complete block design

Block	Genotype	Yield
1	1	5.25
1	2	9
1	3	6.25
1	4	4.5
1	5	5.5
2	1	6.5
2	2	9.5
2	3	6.75
2	4	4.25
2	5	6.5
3	1	4
3	2	6.25
3	3	5.5
3	4	4.5
3	5	5.25
4	1	7
4	2	8.75
4	3	6.75
4	4	5
4	5	6

The above code shows that Genotype was specified as a fixed effect, although Block was specified as a random effect. It is important to point out that both effects were converted to factors. The output of this fitted model is given below.

```
> Fitted
Linear mixed model fit by REML ['lmerMod']
Formula: Yield ~ Genotype + (1 | Block)
  Data: Data_RCBD
REML criterion at convergence: 43.1701
Random effects:
 Groups  Name        Std.Dev.
 Block   (Intercept) 0.6922
 Residual            0.6739
Number of obs: 20, groups: Block, 4
Fixed Effects:
(Intercept)  Genotype2  Genotype3  Genotype4  Genotype5
   5.688      2.688      0.625     -1.125      0.125
> ####Extracting design matrix
> X=Fitted@pp$X
> X=X[!duplicated(X), ]
> X
  (Intercept) Genotype2 Genotype3 Genotype4 Genotype5
1         1         0         0         0         0
5         1         1         0         0         0
9         1         0         1         0         0
13        1         0         0         1         0
17        1         0         0         0         1
```

```
> ####Extracting the fixed effects
> Beta=Fitted@beta
> ####Obtaining the BLUEs of genotypes
> BLUEs_Gen=X%*%Beta
> BLUEs_Gen
     [,1]
1   5.6875
5   8.3750
9   6.3125
13  4.5625
17  5.8125
```

The above code shows that the standard deviation of the random effect of blocks was equal to 0.6922, while the standard deviation of the residual was equal to 0.6739. In the part that reads "Fixed Effects" we find the beta coefficient estimates of the fixed effects with which the BLUEs of each of the genotypes can be obtained. With the Fitted@pp$X, the design matrix of fixed effects is extracted as implemented in lmer, and with X[!duplicated(X),] the rows which are duplicated due to blocks being removed. Then, with Fitted@beta, the beta coefficients for the fixed effects are extracted and, finally, with X%*%Beta, the BLUEs of each genotype are obtained. Since the estimable function (Table 2.1) for genotypes (treatments) are $\eta_0 + \tau_i$, the BLUEs of each genotype are computed as $5.688 + 0 = 5.688$ (BLUE of genotype 1), $5.688 + 2.688 = 8.376$ (BLUE of genotype 2), $5.688 + 0.625 = 6.313$ (BLUE of genotype 3), $5.688 - 1.125 = 4.563$ (BLUE of genotype 4), and $5.688 + 0.125 = 5.813$ (BLUE of genotype 5). Since for more complex models, obtaining the BLUEs for genotypes can be quite laborious, these can be obtained directly using the following lines of code:

```
library(lsmeans)
Lsmeans_Gen=lsmeans(Fitted,~ Genotype)
#Lsmeans_Gen
BLUEs_Gen=data.frame(GID=Lsmeans_Gen$Genotype,
          BLUEs=Lsmeans_Gen$lsmean,
          SE_BLUEs=Lsmeans_Gen$SE)
BLUEs_Gen
```

In this way we get

```
> BLUEs_Gen
 GID BLUEs SE_BLUEs
1  1 5.6875 0.4830459
2  2 8.3750 0.4830459
3  3 6.3125 0.4830459
4  4 4.5625 0.4830459
5  5 5.8125 0.4830459
```

Predictable functions These are linear combinations of the fixed and random effects, $\boldsymbol{K}^{\mathrm{T}}\boldsymbol{\beta} + \boldsymbol{M}^{\mathrm{T}}\boldsymbol{u}$, that is, they can be formed from linear combinations of the conditional means: $\boldsymbol{K}^{\mathrm{T}}\boldsymbol{\beta} + \boldsymbol{M}^{\mathrm{T}}\boldsymbol{u}$ is a predictable function if $\boldsymbol{K}^{\mathrm{T}}\boldsymbol{\beta}$ is estimable. The inference based on these predictable functions is referred to as narrow inference, which, unlike broad inference, has random effects such as additional terms and limits the attention to a group of the sampled random levels (Littell et al. 2006). Then, replacing the estimates obtained from the mixed model equation (2.2) in predictable functions, $\boldsymbol{K}^{\mathrm{T}}\widehat{\boldsymbol{\beta}} + \boldsymbol{M}^{\mathrm{T}}\widehat{\boldsymbol{u}}$ results in the best linear unbiased predictors (BLUPs) of the corresponding predictable function. From a theoretical point of view, BLUP is expected to have a better genotypic predictive accuracy than BLUE, which is important for the selection of new cultivars, or even the genetic values (additive effects) for the selection of progenitors (Piepho et al. 2008). Before BLUP-based selection, selection in crop breeding was based on either simple arithmetic means or BLUEs of genotypes, which can also be calculated in a mixed model context based on fixed genotype effects (Piepho et al. 2008).

BLUP has a long tradition of selection in animal science, but its use in plant breeding is very recent. It is therefore important to provide a clear distinction between the two terms.

Table 2.3 presents the corresponding predictable functions for the same models described in Table 2.1.

It is important to point out that BLUEs and BLUPs are not restricted only to linear mixed models (Eq. 2.1), because they can be obtained in an approximate manner for

Table 2.3 Some predictable functions for the same models given in Table 2.1

Model	Predictor	Predictable function
Complete randomized design (CRD)	$\eta_i = \eta_0 + \tau_i$	None, since there are no random effects
Randomized complete blocks (RCBD)	$\eta_{ij} = \eta_0 + \tau_i + b_j$, $b_j \sim N(0, \sigma_b^2)$	$\eta_0 + \tau_i + b_j$
Regression	$\eta_i = \eta_0 + \sum_{j=1}^{p} x_{ij}\beta_j$	None, since there are no random effects
Split plot design in a CRD	$\eta_{ijk} = \eta_0 + \alpha_i + \alpha(r)_{ik} + \beta_j + (\alpha\beta)_{ij}$; $\beta_j \sim N(0, \sigma_\beta^2)$, $(\alpha\beta)_{ij} \sim N(0, \sigma_{\alpha\beta}^2)$, $\alpha(r)_{ik} \sim N(0, \sigma_{\alpha(r)}^2)$	$\eta_0 + \alpha_i + \beta_j + (\alpha\beta)_{ij}$
Split plot design in RCBD	$\eta_{ijk} = \eta_0 + \alpha_i + r_k + \alpha(r)_{ik} + \beta_j + (\alpha\beta)_{ij}$; $\beta_j \sim N(0, \sigma_\beta^2)$, $(\alpha\beta)_{ij} \sim N(0, \sigma_{\alpha\beta}^2)$, $r_k \sim N(0, \sigma_r^2)$ $\alpha(r)_{ik} \sim N(0, \sigma_{\alpha(r)}^2)$	$\eta_0 + \alpha_i + \beta_j + (\alpha\beta)_{ij}$

Table 2.4 BLUEs for mean response and BLUPs for conditional mean response for different types of response variables

Type of response variable	Distribution	Link function	BLUE	BLUP
Continuous	Normal	Identity	$\boldsymbol{X}\widehat{\boldsymbol{\beta}}$	$\boldsymbol{X}\widehat{\boldsymbol{\beta}}+\boldsymbol{Z}\widehat{\boldsymbol{u}}$
Binary	Binomial	Logit	$\frac{1}{1+\exp\left(-\boldsymbol{X}\widehat{\boldsymbol{\beta}}\right)}$	$\frac{1}{1+\exp\left(-\boldsymbol{X}\widehat{\boldsymbol{\beta}}-\boldsymbol{Z}\widehat{\boldsymbol{u}}\right)}$
Binary	Binomial	Probit	$\Phi\left(\boldsymbol{X}\widehat{\boldsymbol{\beta}}\right)$	$\Phi\left(\boldsymbol{X}\widehat{\boldsymbol{\beta}}+\boldsymbol{Z}\widehat{\boldsymbol{u}}\right)$
Counts	Poisson	Log	$\exp\left(\boldsymbol{X}\widehat{\boldsymbol{\beta}}\right)$	$\exp\left(\boldsymbol{X}\widehat{\boldsymbol{\beta}}+\boldsymbol{Z}\widehat{\boldsymbol{u}}\right)$
Counts	Negative binomial	Log	$\exp\left(\boldsymbol{X}\widehat{\boldsymbol{\beta}}\right)$	$\exp\left(\boldsymbol{X}\widehat{\boldsymbol{\beta}}+\boldsymbol{Z}\widehat{\boldsymbol{u}}\right)$
Continuous proportions	Beta	Logit	$\frac{1}{1+\exp\left(-\boldsymbol{X}\widehat{\boldsymbol{\beta}}\right)}$	$\frac{1}{1+\exp\left(-\boldsymbol{X}\widehat{\boldsymbol{\beta}}-\boldsymbol{Z}\widehat{\boldsymbol{u}}\right)}$
Continuous positives	Gamma	Inverse	$\frac{1}{\boldsymbol{X}\widehat{\boldsymbol{\beta}}}$	$\frac{1}{\boldsymbol{X}\widehat{\boldsymbol{\beta}}+\boldsymbol{Z}\widehat{\boldsymbol{u}}}$

Φ is the cumulative distribution function (CDF) of the standard normal distribution

any fitted generalized linear mixed model using the expression BLUP $= g^{-1}\left(\boldsymbol{X}\widehat{\boldsymbol{\beta}} + \boldsymbol{Z}\widehat{\boldsymbol{u}}\right)$ (Stroup 2012). For example, Table 2.4 provides the BLUEs and BLUPs for some of the most popular response variables under a predictor with fixed and random effects.

In sum, the linear combinations of fixed effects only are called estimable functions and give rise to BLUEs. The solution of mixed model equations produces estimates, or BLUEs, for linear combinations of the form $\boldsymbol{K}^{\mathrm{T}}\boldsymbol{\beta}$. Linear combinations of fixed *and* random effects are called predictable functions. Solving the mixed model equations yields *predictors*, or BLUPs, which are used to obtain BLUPs of linear combinations such as $\boldsymbol{K}^{\mathrm{T}}\boldsymbol{\beta}+\boldsymbol{M}^{\mathrm{T}}\boldsymbol{u}$. The best of both BLUEs and BLUPs means that these estimates have minimum mean square errors (see Searle et al. 2006) for the different meanings of the criteria applied to each one.

Both the BLUEs and BLUPs are not possible to be computed in real applications, since true variance–covariance values of $\boldsymbol{R}$ and $\boldsymbol{\Sigma}$ are required, but we only have access to estimates of these variance–covariance matrices. For this reason, only empirical BLUEs and empirical BLUPs are possible, since we use variance–covariance parameter estimates ($\widehat{\boldsymbol{R}}$ and $\widehat{\boldsymbol{\Sigma}}$) to solve the mixed model equations for $\boldsymbol{\beta}$ and $\mathbf{u}$.

Below, we illustrate the calculation of the BLUPs of genotypes of the data set given in Table 2.2 for which the BLUEs of genotypes were obtained. Using the lmer() function again, but now assuming that the genotype is a random effect, instead of fixed effects we obtained the following output of the fitted model. We can see that the standard deviation of genotypes is 1.3575, the standard deviation of blocks is 0.6922, the standard deviation of residuals is 0.6739, and the intercept is equal to 6.15.

```
> #####BLUP of genotypes
> Fitted2=lmer(Yield~ (1|Genotype) + (1 | Block), Data_RCBD)
> Fitted2
```

```
Linear mixed model fit by REML ['lmerMod']
Formula: Yield ~ (1 | Genotype) + (1 | Block)
  Data: Data_RCBD
REML criterion at convergence: 58.8152
Random effects:
 Groups  Name        Std.Dev.
 Genotype (Intercept) 1.3575
 Block   (Intercept) 0.6922
 Residual            0.6739
Number of obs: 20, groups: Genotype, 5; Block, 4
Fixed Effects:
(Intercept)
    6.15
```

From the fitted model (Fitted2), we now extract the intercept with fixef(Fitted2) and the random effects of genotypes with c(ranef(Fitted2)$Genotype); then we sum up these two terms to get the BLUPs of genotypes that we called BLUP_Gen2, and finally, we calculate the correlation between the BLUPs and BLUEs (obtained above) for the same genotypes. This correlation was equal to one, which shows that we should not expect big differences between the use of BLUPs or BLUEs of genotypes, although there are large amounts of empirical evidence showing that the BLUPs should be preferred over the BLUEs and not always are the same results produced by both.

```
> #####Fixed effect=Intercept######
> Intercept=fixef(Fitted2)
> str(Intercept)
Named num 6.15
 - attr(*, "names")= chr "(Intercept)"
> #####Random effects of genoytpes
> U_ref=c(ranef(Fitted2)$Genotype)
> U_ref
$'(Intercept)'
[1] -0.4356567 2.0958619 0.1530686 -1.4953621 -0.3179116
> #######BLUP of Genotypes#####
> BLUP_Gen2=Intercept+U_ref$'(Intercept)'
> BLUP_Gen2
[1] 5.714343 8.245862 6.303069 4.654638 5.832088
> cor(c(BLUEs_Gen),BLUP_Gen2)
[1] 1
```

2.3 Marker Depuration

First, we will define markers and their importance. Markers are beneficial in the construction of precise genetic relationships, for parental determination and for the identification and mapping of quantitative trait loci (QTL). Between 1970 and 2001, most of the genetic progress in the livestock industry was reached by using pedigree and phenotypic information. However, after the first draft of the human genome project was finished in 2001 (The International SNP Map Working Group 2001), the cost of genotyping using single nucleotide polymorphisms (SNPs) started to decrease considerably, and now its cost is at least 1000 times lower. For this reason, Stonecking (2001) points out that SNPs have become the bread and butter of DNA sequence variation and are essential in determining the genetic potential of livestock and plant breeding.

However, it is also important to point out that other types of DNA markers have been discovered, such as restriction fragment length polymorphisms (RFLP), simple sequence repeat (SSR), Diversity Arrays Technology (DArT), simple sequence length polymorphisms (SSLP), amplified fragment length polymorphisms (AFLP), etc. However, SNPs have become the main markers used to detect DNA variation for some of the following reasons: (a) SNPs are abundant and found throughout the entire genome, in intragenic and extragenic regions (Schork et al. 2000), (b) they represent the most common genetic variants, (c) the location in the DNA: they are found in introns, exons, promoters, enhancers, or intergenic regions, (d) they are easily evaluated by automated means, (e) many of them have direct repercussions on traits of interest in plant and animals, (f) they are generally biallelic, and (g) they are now cheap and easy to genotype.

It is important to remember that DNA (deoxyribonucleic acid) is organized in pairs of chromosomes, each inherited from one of the parents. The diversity found among organisms is a result of variations in DNA sequences and of environmental effects. Genetic variation is substantial and each individual of a species, with the exception of monozygotic twins, possesses a unique DNA sequence. DNA variations are mutations resulting from the substitution of single nucleotides (single nucleotide polymorphisms—SNPs), the insertion or deletion of DNA fragments of various lengths (from a single to several thousand nucleotides), or the duplication or inversion of DNA fragments (Marsjan and Oldenbroek 2007). For this reason, the genome is composed of four different nucleotides (A, C, T, and G). Next, we provide two important definitions that are keys to understanding how markers are used in genomic selection.

Genetic markers A genetic marker is a gene or DNA sequence with a known location on a chromosome and is generally used to identify individuals, which is why it is a powerful tool to explore genetic diversity. It can be described as a variation that may arise due to a mutation or alteration in the genomic loci that can be observed. A genetic marker may be a short DNA sequence, such as a sequence surrounding a single base-pair change (single nucleotide polymorphism, SNP, see Fig. 2.1), or a long one, such as mini- and microsatellites. Molecular markers can be used in molecular biology and biotechnology to identify a particular DNA sequence in a pool of unknown DNA. For example, DNA is used to search for useful genes, and also for marker-assisted selection, paternity testing, and food traceability. In GS, genetic markers measured across the genome are used to measure genomic similarities between individuals; in theory, this can be more precise than pedigree information.

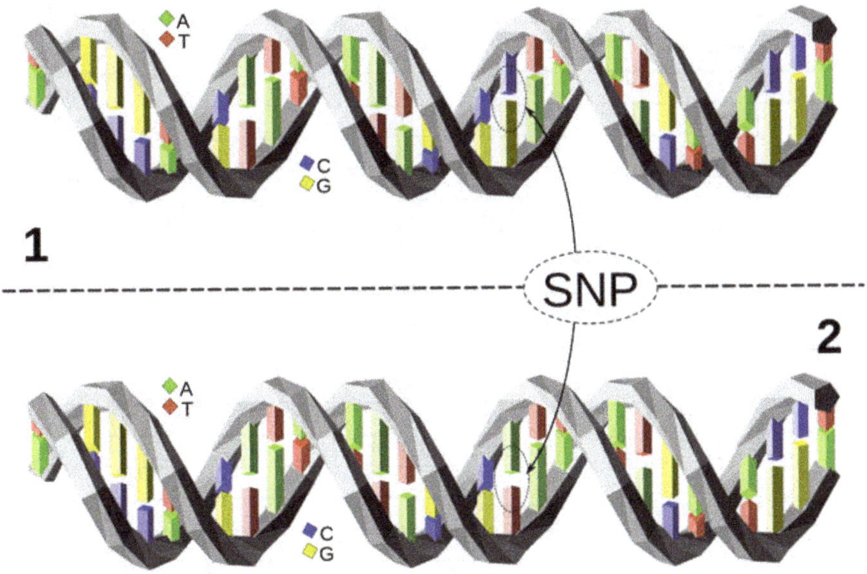

Fig. 2.1 The upper DNA molecule differs from the lower DNA molecule at a single base-pair location (a C/A polymorphism)

Markers can be used to estimate the proportion of chromosome segments shared by individuals, including the identification of genes that are identical by state (IBS). It is important to point out that probabilities generated from pedigree (A matrix) are discrete between close relatives. For example, full sibs share 0.5 of alleles (genome) that are identical by descent, that is, that are inherited from a common ancestor. A simple nucleotide polymorphism (SNP) is a widely used marker.

As an initial descriptive analysis of this type of data that could help delve into some characteristics of the structure of the genotypes data, we can compute the frequency genotypes in our data, basically the proportion of each genotype, and can help us as description of the gene variation. Rather, a more common description of the genetic variation is the allele frequency, which is the proportion of each allele present in the population, and when the individual is a diploid, its contribution of the proportion is of two alleles for each gene (Griffiths et al. 2005). In a diploid population with two alleles, A and a, if n_{A}, n_{Aa}, and n_{a} are the frequencies of the three genotypes (AA, Aa, aa) in a sample of n individuals, the frequency of alleles A and a are given by $\frac{2n_{\mathrm{A}}+n_{\mathrm{Aa}}}{2n}=\frac{n_{\mathrm{A}}}{n}+\frac{n_{\mathrm{Aa}}}{2n}$ and $\frac{2n_{\mathrm{a}}+n_{\mathrm{Aa}}}{2n}=\frac{n_{\mathrm{a}}}{n}+\frac{n_{\mathrm{Aa}}}{2n}$, respectively. In this case, the allele with the lower frequency is called the minor allele and the other, the major allele.

To illustrate marker recoding and depuration, consider the following example of eight plants genotyped for seven SNPs.

With the information in Table 2.5, we first proceed to find the minor (less common) and major (most common) alleles of each marker. In marker 1 (SNP1),

Table 2.5 Marker information for eight plants and seven SNPs denoted as SNP1, . . ., SNP7

Plant	SNP1	SNP2	SNP3	SNP4	SNP5	SNP6	SNP7
1	C_G	C_C	T_T	G_T	G_G	T_C	G_G
2	C_G	C_G	A_A	G_T	C_C	C_C	G_C
3	C_C	?_?	T_A	G_T	G_C	T_C	?_?
4	G_G	?_?	T_T	T_T	G_C	T_C	G_C
5	C_C	C_G	T_A	G_T	G_C	T_T	C_C
6	?_?	C_G	T_A	G_T	C_C	?_?	G_G
7	C_G	C_C	T_A	G_G	G_C	C_C	G_C
8	C_G	C_C	T_A	G_G	G_C	T_C	G_G

?_? denotes a missing genotype

Table 2.6 Minor allele (minorAllele) and major allele (majorAllele) for each marker

Marker	minorAllele	majorAllele
SNP1	G	C
SNP2	G	C
SNP3	A	T
SNP4	T	G
SNP5	G	C
SNP6	T	C
SNP7	C	G

we can see that the minor allele is G, since it only appears in G_G in one out of the eight plants; the major allele is C, since C_C appears in two out of eight plants. In SNP2, the minor allele is G since it does not appear in any of the eight plants, whereas C_C appears in three out of eight plants. Due to this, C is the major allele. In SNP3, the minor and major alleles are A (with A_A observed in 1/8) and T (with T_T observed in 2/8), respectively. Using this logic, Table 2.6 shows the minor and major alleles of each of the markers.

Once we have this information (minor and major alleles), it can be used to fit additive effects models, but it is first necessary to recode the marker information following the rules below:

$$x = \begin{cases} 0 \text{ if the SNP is homozygous for the major allele} \\ 1 \text{ if the SNP is heterozygous} \\ 2 \text{ if the SNP is homozygous for the other allele} \end{cases}$$

The recoded information in terms of additive effects is given in Table 2.7, where the recording is now in terms of 0, 1, and 2, following the above rules. In turn, the missing genotypes are recoded as NA.

Next, we show the minor allele frequency (MAF), the frequency of NAs and the frequency of heterozygotes genotypes (freHetero) for each marker. The frequency allele can be computed from the frequency of genotypes as described earlier and from there, the minor allele and its frequency can be deduced, but once the marker

Table 2.7 Marker information recoded as 0, 1, and 2, for eight plants and seven SNPs denoted as SNP1, ..., SNP7

Plant	SNP1	SNP2	SNP3	SNP4	SNP5	SNP6	SNP7
1	1	0	0	1	2	1	0
2	1	1	2	1	0	0	1
3	0	NA	1	1	1	1	NA
4	2	NA	0	2	1	1	1
5	0	1	1	1	1	2	2
6	NA	1	1	1	0	NA	0
7	1	0	1	0	1	0	1
8	1	0	1	0	1	1	0

NA denotes a missing genotype

Table 2.8 Minor allele frequency (MAF), frequency of NAs (freqNA), and frequency of heterogeneous (freqHetero) are reported for each marker

Marker	minorAllele	majorAllele	MAF	freqNA	freqHetero
SNP1	G	C	0.429	0.125	0.571
SNP2	G	C	0.250	0.250	0.500
SNP3	A	T	0.438	0.000	0.625
SNP4	G	T	0.438	0.000	0.625
SNP5	G	C	0.438	0.000	0.625
SNP6	T	C	0.429	0.125	0.571
SNP7	C	G	0.357	0.125	0.429

information is coded as Table 2.7, in an equivalent way, the MAF also can be calculated as the mean of each column in Table 2.7 divided by 2 without taking into account the missing values, that is, as $\text{MAF} = \left(\sum_{i=1}^{n_c} x_i^*\right)/(2n_c)$, where $x_i^*, i = 1, \ldots, n_c$ are the coded genotyped values for non-missing individual values for a marker. For example, the MAF of marker SNP2 is equal to $\text{MAF}_{\text{SNP2}} = \frac{0+1+1+1+0+0}{2(6)} = \frac{3}{12} = 0.25$. See Table 2.8 for the MAF of the remaining markers, where the frequency of NAs (freqNA) is also reported for each marker, which corresponds to the number of missing values (NAs) in each column divided by the number of individuals (8). Furthermore, the frequency of heterogeneous genotypes (freqHetero) was calculated (ratio of the summation of only the ones divided by the non-missing values).

With the data formatted in this way, we are ready to compute a genomic relationship matrix or matrix of realized genetic similarities among all pairs of individuals.

It is important to point out that the recoding process for values of 0, 1, and 2 can be performed automatically using the library synbreed, but we first need to load the complete information onto R of Table 2.5. Then, for recording and imputing, we used methods available from this library; the code used for these tasks is given in Appendix 1. The output is explained below.

First, we called the library synbreed and then we loaded the marker information contained in a file called MarkersToy.csv, which is saved in an object called snp7. When all the marker information is printed, we can clearly see that it corresponds to the information given in Table 2.5, although without the first column.

```
> library(synbreed)
> ####Loading the marker information
> snp7 <- read.csv("MarkersToy.csv",header=T)
> snp7=snp7[,-1]
> snp7
  SNP1 SNP2 SNP3 SNP4 SNP5 SNP6 SNP7
1 C_G  C_C  T_T  G_T  G_G  T_C  G_G
2 C_G  C_G  A_A  G_T  C_C  C_C  G_C
3 C_C  ?_?  T_A  G_T  G_C  T_C  ?_?
4 G_G  ?_?  T_T  T_T  G_C  T_C  G_C
5 C_C  C_G  T_A  G_T  G_C  T_T  C_C
6 ?_?  C_G  T_A  G_T  C_C  ?_?  G_G
7 C_G  C_C  T_A  G_G  G_C  C_C  G_C
8 C_G  C_C  T_A  G_G  G_C  T_C  G_G
```

Next, we rename the rows of the object snp7 (matrix of marker information) coded with values of ID1 to ID8. We then select the position in this matrix of values equal to ?_?, followed by these values being replaced with NA. Finally, the snp7 object is printed again and we can see that the values of ?_? were replaced by NAs.

```
> ####Set names for individuals
> rownames(snp7) <- paste("ID",1:8,sep="")
> pos.NA=which(snp7=="?_?", arr.ind=TRUE)
> snp7[pos.NA]=NA
> snp7
    SNP1 SNP2 SNP3 SNP4 SNP5 SNP6 SNP7
ID1 C_G  C_C  T_T  G_T  G_G  T_C  G_G
ID2 C_G  C_G  A_A  G_T  C_C  C_C  G_C
ID3 C_C  NA   T_A  G_T  G_C  T_C  NA
ID4 G_G  NA   T_T  T_T  G_C  T_C  G_C
ID5 C_C  C_G  T_A  G_T  G_C  T_T  C_C
ID6 NA   C_G  T_A  G_T  C_C  NA   G_G
ID7 C_G  C_C  T_A  G_G  G_C  C_C  G_C
ID8 C_G  C_C  T_A  G_G  G_C  T_C  G_G
```

Later, the object with the marker information, snp7, is transformed into an object of class gpData.

```
> ####Creating an object of class 'gpData'
> gp <- create.gpData(geno=snp7)
```

Using the function codeGeno() and giving the marker object, gp, as an input, we recode the marker information to values of 0, 1, and 2. The recoded marker information is then extracted and printed, and here we can see that the recoding performed with this library is exactly equal to what we obtained manually and presented in Table 2.7.

```
> ####Recoding to 0, 1, and 2 values the genotypic data
> gp.coded <-codeGeno(gp)
> Geno_Recoded=gp.coded$geno
> Geno_Recoded
     SNP1 SNP2 SNP3 SNP4 SNP5  SNP6 SNP7
ID1  1    0    0    1    2     1    0
ID2  1    1    2    1    0     0    1
ID3  0    NA   1    1    1     1    NA
ID4  2    NA   0    2    1     1    1
ID5  0    1    1    1    1     2    2
ID6  NA   1    1    1    0     NA   0
ID7  1    0    1    0    1     0    1
ID8  1    0    1    0    1     1    0
```

Finally, we once again use the function codeGeno(), but we also add input=T and inpute.type="random", which will recode the object gp to values 0, 1, and 2, as done previously. However, it will also input the missing cells with NAs using the random method of imputation. Finally, the matrix of markers coded with values 0, 1, and 2 is presented, but with the NAs inputted with values of 0, 1, and 2 using the random method. It is important to point out that this library has other imputation options such as "family," "beagle," "beagleAfterFamily," "beagleNoRand," "beagleAfterFamilyNoRand," and "fix," but the technical details of each imputation method go beyond the scope of this book.

```
> ####Recoding to values of 0, 1, and 2 the genotypic data and inputting
> Imputed_Geno<-codeGeno(gp,impute=T,impute.type="random")
> Imputed_Geno$geno
     SNP1 SNP2 SNP3 SNP4 SNP5 SNP6 SNP7
ID1  1    0    0    1    2    1    0
ID2  1    1    2    1    0    0    1
ID3  0    0    1    1    1    1    2
ID4  2    0    0    2    1    1    1
ID5  0    1    1    1    1    2    2
ID6  2    1    1    1    0    0    0
ID7  1    0    1    0    1    0    1
ID8  1    0    1    0    1    1    0
```

2.4 Methods to Compute the Genomic Relationship Matrix

The three methods described here to calculate the genomic relationship matrix (GRM) are based on VanRaden's (2008) paper "Efficient methods to compute genomic predictions" where more theoretical support for each of these methods can be found. We assume that we have a matrix of markers of order $J \times p$, where J denotes the number of lines and p the number of markers, and that this matrix does not contain missing values and is coded as 0, 1, and 2, or −1, 0, and 1 to refer homozygotes major allele, heterozygous, and homozygous minor allele, respectively. Note that the last codification is related to the first by the relation $\boldsymbol{X}_2 = \boldsymbol{X} + \mathbf{1}_J\mathbf{1}_p^{\mathrm{T}}$, where $\boldsymbol{X}_2$ is a matrix of markers information coded in terms of −1, 0, and 1, while $\boldsymbol{X}$ is the coded marker information in terms of 0, 1, and 2, and $\mathbf{1}_q$ is the column vector of dimension q with ones in all its entries.

Method 1. This method calculates the GRM as

$$\boldsymbol{G} = \frac{1}{p}\boldsymbol{X}\boldsymbol{X}^{\mathrm{T}},$$

where $\boldsymbol{X}$ is the matrix of marker genotypes of dimensions $J \times p$. When the marker information is coded as −1, 0, and 1 as described before, the diagonal terms of $p\boldsymbol{G}$ count the number of homozygous loci for each line, and the off-diagonal of $p\boldsymbol{G}$ is a measure of the number of alleles shared by two lines (VanRaden 2008). To illustrate how to calculate the GRM under this method, we will use the matrix of marker genotypes obtained in the previous section with Imputed_Geno$geno, which in the matrix format is equal to

$$\boldsymbol{X} = \begin{bmatrix} 1 & 0 & 0 & 1 & 2 & 1 & 0 \\ 1 & 1 & 2 & 1 & 0 & 0 & 1 \\ 0 & 0 & 1 & 1 & 1 & 1 & 2 \\ 2 & 0 & 0 & 2 & 1 & 1 & 1 \\ 0 & 1 & 1 & 1 & 1 & 2 & 2 \\ 2 & 1 & 1 & 1 & 0 & 0 & 0 \\ 1 & 0 & 1 & 0 & 1 & 0 & 1 \\ 1 & 0 & 1 & 0 & 1 & 1 & 0 \end{bmatrix}$$

Then, using the R code, we calculate the GRM under this first method as

```
##Computing the genomic relationship matrix–Method1
> G_M1=tcrossprod(X)/dim(X) [2]
> G_M1
      ID1   ID2   ID3   ID4   ID5   ID6   ID7   ID8
ID1 0.875 0.250 0.500 0.875 0.625 0.375 0.375 0.500
ID2 0.250 1.000 0.625 0.625 0.750 0.750 0.500 0.375
ID3 0.500 0.625 1.000 0.750 1.125 0.250 0.500 0.375
ID4 0.875 0.625 0.750 1.375 0.875 0.750 0.500 0.500
ID5 0.625 0.750 1.125 0.875 1.500 0.375 0.500 0.500
ID6 0.375 0.750 0.250 0.750 0.375 0.875 0.375 0.375
ID7 0.375 0.500 0.500 0.500 0.500 0.375 0.500 0.375
ID8 0.500 0.375 0.375 0.500 0.500 0.375 0.375 0.500
```

Method 2. In this method the GRM is similar to method 1, but first each marker is centered by twice the minor allele frequency:

$$G = \frac{(X - \mu_E)(X - \mu_E)^T}{2\sum_{j=1}^{p} p_j(1 - p_j)},$$

where p_j is the minor allele frequency (MAF) of SNP $j = 1, \ldots, p$ and μ_E is the expected value of matrix X under the Hardy–Weinberg equilibrium (Griffiths et al. 2005) from estimates of allelic frequencies, that is, $\mu_E = \mathbf{1}_J[2p_1, \ldots, 2p_p]$. Term $2\sum_{j=1}^{p} p_j(1 - p_j)$ is the sum of the variance estimates of each marker and makes GRM analogous to the numerator relationship matrix (VanRaden 2008).

Now, using the following R code, we calculate the GRM under method 2 as

```
##Computing the genomic relationship matrix—Method2
> phat=colMeans(X)/2#Minor allele frequency
> phat
 SNP1   SNP2   SNP3   SNP4   SNP5   SNP6   SNP7
 0.5000 0.1875 0.4375 0.4375 0.4375 0.3750 0.4375
```

```
> X2=scale(X,center=TRUE,scale=FALSE)
> k=2*sum(phat*(1-phat))
> G_M2=tcrossprod(X2)/k
```

```
> round(G_M2,3)
       ID1    ID2    ID3    ID4    ID5    ID6    ID7    ID8
ID1  0.930 -0.766 -0.227  0.352 -0.265 -0.227 -0.072  0.275
ID2 -0.766  0.930 -0.072 -0.419 -0.111  0.545  0.082 -0.188
ID3 -0.227 -0.072  0.776 -0.188  0.737 -0.766  0.005 -0.265
ID4  0.352 -0.419 -0.188  1.007 -0.227  0.120 -0.342 -0.304
ID5 -0.265 -0.111  0.737 -0.227  1.316 -0.805 -0.342 -0.304
ID6 -0.227  0.545 -0.766  0.120 -0.805  1.084  0.005  0.043
ID7 -0.072  0.082  0.005 -0.342 -0.342  0.005  0.467  0.198
ID8  0.275 -0.188 -0.265 -0.304 -0.304  0.043  0.198  0.545
```

Method 3. Under this method, the GRM should be calculated as

$$G = \frac{ZZ^T}{p},$$

where Z is the matrix of scaled SNP codes and p is the number of SNPs, that is $z_{ij} = (x_{ij} - 2p_j)/\sqrt{2p_j(1 - p_j)}$.

Finally, for this third method, the GRM can be calculated as

```
> ##Computing the genomic relationship matrix–Method3
> X3=scale(X,center=TRUE,scale=TRUE)
> G_M3=tcrossprod(X3)/ncol(X3)
> round(G_M3,3)
      ID1    ID2    ID3    ID4    ID5    ID6    ID7    ID8
ID1  0.962 -0.880 -0.093  0.435 -0.221 -0.397 -0.028  0.223
ID2 -0.880  1.084 -0.133 -0.507 -0.014  0.667  0.012 -0.228
ID3 -0.093 -0.133  0.619 -0.112  0.490 -0.658  0.023 -0.136
ID4  0.435 -0.507 -0.112  1.058 -0.241  0.022 -0.350 -0.305
ID5 -0.221 -0.014  0.490 -0.241  1.181 -0.539 -0.391 -0.265
ID6 -0.397  0.667 -0.658  0.022 -0.539  1.053 -0.057 -0.092
ID7 -0.028  0.012  0.023 -0.350 -0.391 -0.057  0.516  0.276
ID8  0.223 -0.228 -0.136 -0.305 -0.265 -0.092  0.276  0.527
```

2.5 Genomic Breeding Values and Their Estimation

In plant and animal breeding, it is a common practice to rank and select individuals (plants or animals) based on their true breeding values (TBVs), also called additive genetic values. However, since we cannot see genes and breeding values, this task is not straightforward, and it is therefore estimated indirectly using observed phenotypes. The estimated values are called estimated breeding values (EBVs), which means that TBV is a latent variable that is only approximated using the observable variable (phenotype).

When the TBVs are used, the genetic change is expected to be larger than when the EBVs are used, but this difference is small when the EBVs are accurately estimated. EBVs reflect the true genetic potential or true genetic transmitting ability of individuals (plants or animals). Traditionally, they are estimated based on the performance records of their parents, sibs, progenies, and their own after correcting for various environmental factors such as management, season, age, etc. When parents are selected based on their breeding values with high reliability, a faster genetic progress is expected in the resulting population. For this reason, the process of estimating breeding values is of paramount importance in any breeding program.

There are several methods to estimate genomic estimated breeding values (GEBVs), but first we will describe the best linear unbiased predictor (BLUP) method. When using the BLUP method to estimate the GEBVs, we need to use the mixed model equations (2.2) described above to estimate BLUEs and BLUPs. Using this equation (2.2) but depending on the form taken by the matrices $\boldsymbol{Z}$ and $\boldsymbol{\Sigma}$, we can end up with the GBLUP method or the SNP-BLUP method to estimate the breeding values. First, we explain the GBLUP method, where we substitute $\boldsymbol{Z}$ and $\boldsymbol{\Sigma}$ matrices for the incidence matrix of genotypes and genomic relationship matrix (GRM) derived from allele frequencies calculated with one of the methods of

VanRaden (2008) given in Sect. 2.4. Under this GBLUP method, the GEBV can be obtained as the solution $\widehat{\boldsymbol{u}}$ of the mixed model equation:

$$\begin{pmatrix} \widehat{\boldsymbol{\beta}} \\ \widehat{\boldsymbol{u}} \end{pmatrix} = \begin{pmatrix} \boldsymbol{X}^{\mathrm{T}}\boldsymbol{R}^{-1}\boldsymbol{X} & \boldsymbol{X}^{\mathrm{T}}\boldsymbol{R}^{-1}\mathbf{1} \\ \mathbf{1}^{\mathrm{T}}\boldsymbol{R}^{-1}\boldsymbol{X} & \mathbf{1}^{\mathrm{T}}\boldsymbol{R}^{-1}\mathbf{1} + \sigma_g^{-2}\boldsymbol{G}^{-1} \end{pmatrix}^{-1} \begin{pmatrix} \boldsymbol{X}^{\mathrm{T}}\boldsymbol{R}^{-1}\boldsymbol{y} \\ \mathbf{1}^{\mathrm{T}}\boldsymbol{R}^{-1}\boldsymbol{y} \end{pmatrix}, \tag{2.3}$$

where $\boldsymbol{Z}$ was replaced by $\boldsymbol{Z} = \mathbf{1}$ and $\boldsymbol{\Sigma}$ by $\sigma_g^2\boldsymbol{G}$, the genomic relationship matrix that was calculated with some of the methods described in Sect. 2.4 and the genomic variance component (σ_g^2) should be estimated. We ended up with the system of equations given in Eq. (2.3), since the model used is $\boldsymbol{Y} = \boldsymbol{X\beta} + \boldsymbol{u} + \boldsymbol{\varepsilon}$, with $\boldsymbol{u} \sim N\left(\mathbf{0}, \sigma_g^2\boldsymbol{G}\right)$, where σ_g^2 is the genomic variance component, $\boldsymbol{G}$ is a GRM of dimension $q \times q$ calculated using any of the three methods given in Sect. 2.4, and the other terms are exactly as in Eq. (2.2). One of the greatest advantages of using the GBLUP method to obtain GEBV is that the dimensionality of the design matrices is, at most, equal to the number of lines under study. On the other hand, under the SNP-BLUP, we substitute the $\boldsymbol{Z}$ and $\boldsymbol{\Sigma}$ matrices in Eq. (2.2) with $\boldsymbol{M}$ (scaled marker information matrix of order $n \times p$) and $\sigma_M^2\boldsymbol{I}$, respectively. Under this SNP-BLUP method, the mixed model equation is equal to

$$\begin{pmatrix} \widehat{\boldsymbol{\beta}} \\ \widehat{\boldsymbol{u}} \end{pmatrix} = \begin{pmatrix} \boldsymbol{X}^{\mathrm{T}}\boldsymbol{R}^{-1}\boldsymbol{X} & \boldsymbol{X}^{\mathrm{T}}\boldsymbol{R}^{-1}\boldsymbol{M} \\ \boldsymbol{M}^{\mathrm{T}}\boldsymbol{R}^{-1}\boldsymbol{X} & \boldsymbol{M}^{\mathrm{T}}\boldsymbol{R}^{-1}\boldsymbol{M} + \sigma_M^{-2}\boldsymbol{I}^{-1} \end{pmatrix}^{-1} \begin{pmatrix} \boldsymbol{X}^{\mathrm{T}}\boldsymbol{R}^{-1}\boldsymbol{y} \\ \boldsymbol{M}^{\mathrm{T}}\boldsymbol{R}^{-1}\boldsymbol{y} \end{pmatrix} \tag{2.4}$$

Under this mixed model equation, $\boldsymbol{u}$ is now the random effects of markers, and therefore, to obtain the GEBV, we use the estimates of marker effects ($\widehat{\boldsymbol{u}}$) and GEBV $= \boldsymbol{M}\widehat{\boldsymbol{u}}$, since now the model used is $\boldsymbol{Y} = \boldsymbol{X\beta} + \boldsymbol{Mu} + \boldsymbol{\varepsilon}$, with $\boldsymbol{u} \sim N\left(\mathbf{0}, \sigma_M^2\boldsymbol{I}\right)$.

We then illustrate how to estimate the GEBV under both BLUP methods. For this purpose, we provide, in Table 2.9, a data set with eight lines (evaluated in two environments) for grain yield, for which, in turn, we use the seven markers imputed in Sect. 2.3.

Below are the data of Table 2.9 that were saved in the data.for.GEBV.csv file. Library rrBLUP was used to estimate the GEBV using the mixed model equations.

```
> library(rrBLUP)
> data=read.csv("data.for.GEBV.csv")
> data
  X Env Lines y      SNP1 SNP2 SNP3 SNP4 SNP5 SNP6 SNP7
1 1 E1  L1    5.215 1    0    0    1    2    1    0
2 2 E1  L2    4.998 1    1    2    1    0    0    1
3 3 E1  L3    5.284 0    0    1    1    1    1    2
4 4 E1  L4    5.157 2    0    0    2    1    1    1
5 5 E2  L5    6.601 0    1    1    1    1    2    2
6 6 E2  L6    5.735 2    1    1    1    0    0    0
7 7 E2  L7    5.565 1    0    1    0    1    0    1
8 8 E2  L8    5.829 1    0    1    0    1    1    0
```

Table 2.9 Grain yield (y) of eight lines in two environments

Env	Lines	y	SNP1	SNP2	SNP3	SNP4	SNP5	SNP6	SNP7
E1	L1	5.215	1	0	0	1	2	1	0
E1	L2	4.998	1	1	2	1	0	0	1
E1	L3	5.284	0	0	1	1	1	1	2
E1	L4	5.157	2	0	0	2	1	1	1
E2	L5	6.601	0	1	1	1	1	2	2
E2	L6	5.735	2	1	1	1	0	0	0
E2	L7	5.565	1	0	1	0	1	0	1
E2	L8	5.829	1	0	1	0	1	1	0

In matrix M, only the columns corresponding to marker information are selected. This information is scaled by column using the scale command of R, and the scaled markers are saved in the MS matrix; the GRM is calculated with this information using method 3 in Sect. 2.4. Here we obtained the design matrix for environments and lines, and we also added the genomic information to the design matrix of lines by post-multiplying the design matrix of lines by the Cholesky decomposition of the GRM, which also can be used as an alternative way to obtain the breeding values. This is because the GBLUP model (2.1) can be expressed equivalently as follows:

$$\boldsymbol{Y} = \boldsymbol{X\beta} + \boldsymbol{Z}^{*}\boldsymbol{u}^{*} + \boldsymbol{\varepsilon},$$

where now $\boldsymbol{Z}^{*} = \boldsymbol{Z}\boldsymbol{L}^{\mathrm{T}}$, $\boldsymbol{G} = \boldsymbol{L}^{\mathrm{T}}\boldsymbol{L}$ is the Cholesky decomposition of $\boldsymbol{G}$, and $\boldsymbol{u}^{*}$ is a random vector with distribution $N\left(\boldsymbol{0}, \sigma_g^2\boldsymbol{I}_q\right)$.

```
> M=data[,5:11]
> MS=scale(M) #Scales matrix of markers
> G=MS%*%t(MS)/ncol(MS) #Genomic relationship matrix method 3
> X_E=model.matrix(~0+Env,data=data) #Matrix design of environments
> X_L1=model.matrix(~0+Lines,data=data)#Matrix design of lines
> L=chol(G) #Cholesky decomposition of G
> X_L=X_L1%*%t(L)#Modified matrix design of lines, Z*
```

Then, using the mixed.solve() function of the rrBLUP package, providing as input the response variable (y), the X_L1 matrix of lines and the GRM matrix, $\boldsymbol{G}$, we solved the mixed model equation (2.3) and obtained the GEBVs, which are extracted using fm1$u.

```
> ########################Solution GBLUP##############################
> y = data$y
> fm1=mixed.solve(y=y, Z=X_L1, K=G, X=X_E, method="REML",
+       bounds=c(1e-09, 1e+09), SE=FALSE, return.Hinv=FALSE)
> fm1$u
[1] 0.08371533 -0.13384553 0.15177641 0.02540245 0.63649420
-0.22942366 -0.39877555 -0.13534366
###################Alternative solution GBLUP#######################
```

```
> fm1a=mixed.solve(y=y, Z=X_L, K=diag(dim(G)[1]), X=X_E,
method="REML",
         bounds=c(1e-09, 1e+09), SE=FALSE, return.Hinv=FALSE)
>#GEBV
>X_L%*%fm1a$u
```

Shown below is the code to obtain the GEBV, but using the SNP-BLUP method and also using the mixed.solve() function. However, instead of giving the GRM as input, this is given by **Z**, which is the MS matrix, containing the matrix of the markers scaled. Now the random effects of markers are extracted with fm2$u, and to obtain the GEBV, these random effects are pre-multiplied by the scaled design matrix of markers:

```
> #######################Solution SNP-BLUP########################
> fm2=mixed.solve(y=y, Z=MS, X=X_E, method="REML",
+       bounds=c(1e-09, 1e+09), SE=FALSE, return.Hinv=FALSE)
> fm2$u
SNP1         SNP2         SNP3          SNP4          SNP5         SNP6         SNP7
-0.121863421 0.115618300 -0.040677892 0.083526359 0.016201822 0.184761140 -0.001827558
> beta_Mar=fm2$u
> GEBV=c(MS%*%beta_Mar)
> GEBV
[1] 0.08373722 -0.13385647 0.15181358 0.02538892 0.63650107 -0.22940375 -0.39883111
-0.13534946
```

This shows that both methods (GBLUP and SNP-BLUP), give exactly the same breeding value estimates. However, although the two methods are specified in almost the same way and give similar estimates, which, based on all the SNPs (SNP-BLUP), required more computational time, this difference will be more notable for larger data sets that containing larger number of markers. Due to this, in these situations we can choose the GBLUP method.

Some advantages of using the Henderson equation to obtain the GEBV are

(a) It fits nicely into existing BLUP software and into existing theory.
(b) It provides measures of accuracy from the inverse of the LHS (Linear Henderson system).
(c) It accommodates all individuals (plants or animals).

However, it has some inconveniences, such as

(a) It can't easily accommodate major genes (unless using weights in the construction of $\boldsymbol{G}$).
(b) Computation of $\boldsymbol{G}$ and inversion might be challenging.

It is important to point out that the GEBV can be obtained using other estimation methods such as Bayesian methods. Bayesian methods for GS are explained in detail in upcoming chapters, but here we will illustrate the use of the BGLR package to estimate the GBLUP Bayesian method and the SNP-BLUP Bayesian method.

We first provide the code for the GBLUP Bayesian method. The predictor ETA is a list and has two sub-list components: the first, for the effects of environments for which a FIXED model (model="FIXED") is used that uses a prior non-informative for each beta coefficient. The second component is an RKHS model to specify the distribution of the random effects of lines that, in general, are of the form $N\left(\mathbf{0}, \sigma_g^2 \boldsymbol{K}\right)$, and in this case, using the GRM, $\boldsymbol{K} = \boldsymbol{G}$. Then the Bayesian GBLUP model is fitted using the function BGLR(), and finally, the GEBVs are obtained with fm1ETAGen$u. The response vector value is specified in the first option of the function BGLR, y=y, and with nIter=20000 and burnIn=10000, the number of iterations to run the involved MCMC method are specified, along with the number of these that will be discarded at the beginning of the MCMC, respectively.

```
> library(BGLR)
> ####################GBLUP-BLUP Bayesian##########################
> ETA=list(Env=list(X=X_E[,-1], model="FIXED"),Gen=list(K=G,
model="RKHS"))
> fm1=BGLR(y=y,ETA=ETA,nIter=20000,burnIn=10000,verbose=F)
> fm1$ETA$Gen$u
[1] 0.02581315 -0.08430838 0.12806091 0.03949064 0.32406895 -0.17367228
[7] -0.16452623 -0.09492676
```

The SNP-BLUP Bayesian GEBV is then fitted, but now also providing the design matrix of environments as input, and the scaled design matrix of markers in the second term. The model now used is a Bayesian Ridge regression (BRR) that gives a normal distribution with mean zero and a common variance component as a prior for each marker effect.

```
> ########################SNP-BLUP Bayesian########################
> ETA=list(Env=list(X=X_E[,-1], model="FIXED"), Gen=list(X=MS,
model="BRR"))
> fm1=BGLR(y=y,ETA=ETA,nIter=20000,burnIn=10000,verbose=F)
> beta_Mar_Bayes=fm1$ETA$Gen$b
>
> GEBV_Bayes=c(MS%*%beta_Mar_Bayes)
> GEBV_Bayes
[1] 0.03187469 -0.09663427 0.12920479 0.04973765 0.34695500
-0.18172760
[7] -0.18217036 -0.09723989
```

Here we can observe that both methods gave GEBVs that are very similar yet slightly different due to the Monte Carlo sampling. However, these Bayesian GEBVs are different from those obtained above using Henderson's mixed model equations, due to the fact that different machineries are used to estimate the GEBVs. Details of the Bayesian methods for GS will be provided in upcoming chapters.

Finally, it is important to point out that the advantage of the GBLUP over the SNP-BLUP is that the system of equations, when fitting the mixed model equations, is of the size of the individuals (lines or animals), which are, most of the time, fewer than the number of markers (SNPs). This advantage is also observed under the Bayesian version, because the design matrix of markers is usually larger than the dimension of the genomic relationship matrix.

2.6 Normalization Methods

This section describes four types of normalization variables (inputs and outputs). In this case, normalization refers to the process of adjusting the different inputs or outputs that were originally measured in different scales to the same scale. It is very important to carry out the normalization process before giving the inputs and outputs for most statistical machine learning algorithms because it helps improve the numerical stability in the estimation process of some algorithms; it is suggested mostly when the inputs or outputs are in different scales. However, it is important to point out that in some statistical machine learning software, the normalization process is done internally, in which case this process does not need to be carried out manually. The five normalization methods we describe next are centering, scaling, standardization, max normalization, and minimax normalization.

Centering This normalization consists of subtracting from each variable (input or output) its mean, μ; this means that the centered values are calculated as

$$X_i^* = X_i - \mu$$

The centered variable X_i^* has a mean of zero.

Scaling This normalization consists of dividing each variable (input or output) by its standard deviation, σ. The scaled values are calculated as

$$X_i^* = \frac{X_i}{\sigma}.$$

The scaled variable X_i^* has unit variance.

Standardization This process of normalization consists of calculating its mean, μ, and standard deviation, σ, for each input or output. The standardized values are then calculated as

$$X_i^* = \frac{X_i - \mu}{\sigma}.$$

This process is carried out for each input or output variable, and this needs to be done with care, since we need to use the corresponding mean and standard deviation of each variable. The output of the standardized score has a mean of zero and a variance of one, which means that most standardized values range between −3.5 and 3.5.

Max normalization This normalization consists of dividing the values of the input or output by the maximum (max) value of this variable, meaning that this score is calculated as

$$X_i^* = \frac{X_i}{\max}$$

This normalization can be useful when there are no negative inputs, which guarantees that the normalized variable will be between 0 and 1.

Minimax normalization To implement this normalization, we first need to calculate the minimum (min) and maximum (max) value for each input or output; then the minimax score is calculated using the following expression:

$$X_i^* = \frac{X_i - \min}{\max - \min}$$

The resulting score of the minimax normalization is between 0 and 1. An inconvenience of this normalization method is that inputs or outputs with long-tail distributions will be dominated by inputs or outputs with uniform distributions.

It is important to point out that the normalization process is not limited to the independent variables (inputs), but can also be used for the dependent variables (outputs) when dealing with multiple outcomes in different scales. However, the normalization process of the dependent variables is not necessary for univariate prediction models or when developing a machine to predict mixed outcomes, because the original scale of the distributions can be losses, for example, to predict two types of outcomes (binary and continuous), to predict three types of outcomes (ordinal, continuous, and count), or to predict four types of outcomes (binary, ordinal, continuous, and count data). On the other hand, when developing a machine to predict four continuous outcomes in different scales (for example, grain yield in tons by hectare, plant height in centimeters, days to maturity in 0 to 120 days, and vitamin content in milligrams), in these cases, normalizing each of the response variables is suggested to avoid the training process from being dominated by the dependent variable with large variability, which implies that the trained machine would be able to predict only this response variable with the highest accuracy. However, in some statistical machine learning models, it is not necessary to normalize the dependent variables because they allow the user to put different weights on each dependent variable to be able to train the model more fairly.

2.7 General Suggestions for Removing or Adding Inputs

The following is a general guide to removing inputs:

(a) Remove an independent variable (input) if it has zero variance, which implies that the input has a single unique value (Kuhn and Johnson 2013).

(b) Remove an independent variable (input) if it has near-zero variance, which implies that the input has very few values.
(c) Remove an independent variable (input) if it is highly correlated with another input variable (nearly perfect correlation), since they are measuring the same underlying information (Kuhn and Johnson 2013). Known as collinearity in statistical machine learning science, this phenomenon is important because in its presence the parameter estimates of some machine learning algorithms (for example, those based on gradient descent) are inflated (not accurately estimated).

These three issues are very common in genomic prediction, since part of the independent variables is marker information and many of them have zero or near-zero variance and other pairs have very high correlations. One of the advantages of removing input information prior to the modeling process is that this reduces the computational resources needed to implement the statistical machine learning algorithm. Also, it is possible to end up with a more parsimonious and interpretable model. Another advantage is that models with less correlated inputs are less prone to unstable parameter estimates, numerical errors, and degraded prediction performance (Kuhn and Johnson 2013).

The following are general rules for the addition of input variables:

(a) Create dummy variables from nominal or categorical inputs.
(b) Manually create a categorical variable from a continuous variable.
(c) Transform the original input variable using a specific transformation.

First, we describe the process of creating dummy variables from categorical (nominal or ordinal) inputs. Transforming categorical inputs into dummy variables is required in most supervised statistical machine learning methods, since providing the original independent variable (not transformed into dummy variables) is incorrect and should be avoided by practitioners of statistical machine learning methods. However, it is important to point out that when the dependent variable is categorical, most statistical machine learning methods do not require it to be transformed into dummy variables. For example, assume that we are studying three genotypes (G1, G2, and G3) in two environments (E1 and E2) and we collected the following grain yield data.

Using the information in Table 2.10, we created the dummy variables for each categorical variable. First, we provide the dummy variables for the environments (Table 2.11).

Next, we provide the dummy variables for the genotypes (Table 2.12).

It is important to point out that in R we can use the model.matrix() to create dummy variables from categorical independent variables. First, we create a data frame called grain.yield with the original data set:

```
grain.yield=data.frame(Environment=c("E1","E1", "E1", "E2", "E2",
"E2"), Genotype=c("G1","G2","G3", "G1","G2","G3"), y=c(5.3,
5.6,5.8,6.5, 6.8,6.9))
```

Table 2.10 Grain yield was evaluated in two environments, and three genotypes were evaluated in each environment

Observation	Environment	Genotype	Grain yield (y)
1	E1	G1	5.3
2	E1	G2	5.6
3	E1	G3	5.8
4	E2	G1	6.5
5	E2	G2	6.8
6	E2	G3	6.9

Table 2.11 Resulting dummy variables for the environments

Observation	Env1	Env2
1	1	0
2	1	0
3	1	0
4	0	1
5	0	1
6	0	1

Table 2.12 Resulting dummy variables for genotypes

Observation	G1	G2	G3
1	1	0	0
2	0	1	0
3	0	0	1
4	1	0	0
5	0	1	0
6	0	0	1

We then print the data

```
> grain.yield
     Environment Genotype  y
  1  E1          G1        5.3
  2  E1          G2        5.6
  3  E1          G3        5.8
  4  E2          G1        6.5
  5  E2          G2        6.8
  6  E2          G3        6.9
```

Next, we create the dummy variables for the categorical variable environment using the model.matrix() function, as

```
ZE=model.matrix(~0+Environment, data=grain.yield)
```

The resulting matrix with the dummy variables of environments is called design matrix of environments and, in this case, it is

```
> ZE
   EnvironmentE1 EnvironmentE2
1        1             0
2        1             0
3        1             0
4        0             1
5        0             1
6        0             1
```

It is important to point out that, if instead of ~0+Environment, we use ~1-+Environment inside the model.matrix() function, we obtain a different form of the design matrix for environments:

```
> ZE
 (Intercept) EnvironmentE2
1      1             0
2      1             0
3      1             0
4      1             1
5      1             1
6      1             1
```

Strictly speaking, both design matrices contain the same information due to the fact that only $C - 1$ dummy variables are needed to capture all the information of the categorical variable, since once we have the information of $C - 1$ dummy variables, we can infer the information of the missing dummy variable. However, the decision to include all dummy variables depends on the selected statistical machine learning algorithm. Under the second version of the design matrix created with the model. matrix() function, in ZE, the column (dummy variable) corresponding to the first environment was not included, but instead an intercept (a column of ones in all rows) was added. This design matrix with an intercept is very important in some statistical machine learning algorithms such as most generalized regression models, neural networks, and deep learning models. In some cases, when this intercept is not included, numerical problems occur in the estimation of the learnable parameters (beta coefficients or weights, intercepts, etc.). The reason is that for each row (observation), these variables all add up to one, and this would provide the same information as the intercept (Kuhn and Johnson 2013). However, when statistical machine learning is not sensitive to not including the intercept (as the first design matrix), using the complete set of dummy variables can help improve model interpretation.

In the same way, the design matrix (dummy variables) for genotype is created using the following R code:

```
ZG=model.matrix(~0+Genotype, data=grain.yield)
```

which provides the following dummy variables that are accommodated in the ZG matrix:

```
> ZG
  GenotypeG1 GenotypeG2 GenotypeG3
1          1          0          0
2          0          1          0
3          0          0          1
4          1          0          0
5          0          1          0
6          0          0          1
```

This design matrix is composed of three columns because there are three categories for the categorical variable (G1, G2, and G3). Each column represents a genotype and a dummy variable was created for each genotype using an indicator variable of 0 (not present in that row) and 1 (present in that row) for each genotype. Now, using model.matrix(~1+Genotype, data=grain.yield), we obtain the design matrix with an intercept, but without the dummy variable for genotype 1:

```
> ZG
 (Intercept) GenotypeG2 GenotypeG3
1          1          0          0
2          1          1          0
3          1          0          1
4          1          0          0
5          1          1          0
6          1          0          1
```

As mentioned earlier, both design matrices for genotypes are valid since they contain the same information, given that in order to capture all the information of a categorical variable, reporting only $C - 1$ dummy variables in the design matrix is enough. However, the choice of one or another depends mostly on the statistical machine learning model to be used. It is also important to point out that the model.matrix() function, when containing the intercept, deletes the dummy variable corresponding to the first level of the categorical variable, but from the statistical point of view, any other dummy variable can be deleted without a loss of information if, and only if, $C - 1$ dummy variables are maintained.

Regarding the second point (b: manually create a categorical variable from a continuous variable), we refer to the process of manually converting a continuous variable to a categorical variable. For example, let us assume that we measured plant height in centimeters and then decided to categorize this response variable into five groups (group 1 if plant height is less than 100 cm, group 2 if plant height is between 100 and 125 cm, group 3 if it is between 125 and 150 cm, group 4 if it is between 150 and 175 cm, and group 5 when plant height is greater than 175 cm). This type of categorization is sometimes required, although we suggest avoiding categorizing continuous outcomes in this way, since a significant amount of information is lost and will affect the prediction performance of the trained model. In addition, it is important to point out that the smaller the number of categories created, the greater the loss of information. Also, some researchers like Austin and Brunner (2004) reported that categorizing continuous inputs increases the rate of false positives (Kuhn and Johnson 2013). However, if researchers can justify that categorization is necessary, they should categorize the continuous input or output using a model framework to be able to do this with more precision.

Regarding the third approach to adding or creating inputs by transforming the original input variable using a specific transformation, we refer to the use of kernels, in which the input variables are transformed in such a way that the transformed input is used in the modeling process and the type of transformation depends on the objective of the study. This type of transformation with kernels will be discussed in detail in upcoming chapters. Also, many times transformations are applied to guarantee normality or any other distributional assumption on the variable of interest.

2.8 Principal Component Analysis as a Compression Method

Principal component analysis (PCA) is a method often used to compress the input data without losing as much information. The PCA works on a rectangular matrix in which the rows represent the observations (n) and the columns, the independent variables (p). The PCA creates linear combinations of the columns of matrix information, $\boldsymbol{X}$, and generates, at most, p linear combinations, called principal components. These linear combinations, or principal components, can be obtained as follows:

$$\mathrm{PC}_1 = \boldsymbol{w}_1\boldsymbol{X} = w_{11}X_1 + w_{12}X_2 + \cdots + w_{1p}X_p$$

$$\cdots$$

$$\mathrm{PC}_p = \boldsymbol{w}_p\boldsymbol{X} = w_{p1}X_1 + w_{p2}X_2 + \cdots + w_{pp}X_p$$

These linear combinations are constructed in such a way that the first principal component, PC_1, captures the largest variance, the second principal component, PC_2, captures the second largest variance, and so on. For this reason, it is expected that few principal components ($k < p$) can explain the largest variability contained in the original rectangular matrix ($\boldsymbol{X}$), which means that with a compressed matrix, $\boldsymbol{X}^*$, we contain most of the variability of the original matrix, but with a significant reduction in the number of columns. In matrix notation, the full principal components are obtained with the following expression:

$$\mathbf{PC} = \boldsymbol{XW},$$

where $\boldsymbol{W}$ is a p-by-p matrix of weights whose columns are the eigenvectors of $\boldsymbol{Q} = \boldsymbol{X}^{\mathrm{T}}\boldsymbol{X}$, that is, we first need to calculate the eigenvalue decomposition of $\boldsymbol{Q}$, which is equal to $\boldsymbol{Q} = \boldsymbol{W}\boldsymbol{\Lambda}\boldsymbol{W}^{\mathrm{T}}$, where $\boldsymbol{W}$ represents the matrix of eigenvectors and $\boldsymbol{\Lambda}$ is a diagonal matrix of order p-by-p containing the eigenvalues. For this reason, if we use $k < p$ principal components, the reduced (compressed) matrix is of order $n \times k$ and is calculated as

$$\boldsymbol{X}^* = \boldsymbol{X}\boldsymbol{W}^*,$$

where $\boldsymbol{W}^*$ contains the same rows of $\boldsymbol{W}$, but only the first k columns instead of the original p columns. The selection of the number of principal components to maintain is critical, and we therefore provide some classical rules for this process:

(a) Select the required principal components to cover a certain amount of variances, such as 80% or 90%.
(b) Order the eigenvalues from highest to lowest, then make a plot of each of the ordered eigenvalues against its position and select as the number of principal components that number from which little variance is gained by retaining additional eigenvalues. This plot is called a scree plot.
(c) Discard those components associated with eigenvalues below a certain level, which is usually set as the average variance. In particular, when working with a correlation matrix built with the input matrix, $\boldsymbol{X}$, the average value of the components is 1, and this rule leads to selecting eigenvalues greater than the unit.

It is important to point out that the principal components can be obtained from a covariance matrix, $\boldsymbol{Q} = \frac{1}{n-1}\boldsymbol{X}^{\mathrm{T}}\boldsymbol{X}$, where each column of $\boldsymbol{X}$ is centered, or from the correlation matrix, $\boldsymbol{Q} = \frac{1}{n-1}\boldsymbol{X}^{\mathrm{T}}\boldsymbol{X}$, where each column of the original matrix of information was standardized. The covariance matrix is used to calculate the principal components when all the independent variables were measured using the same scale, but if they were measured in different scales, we recommend calculating the principal components with the correlation matrix, which agrees with the normalization methods that are suggested when the independent variables were measured in different scales.

Assume that we measured 15 observations (lines) and five independent variables, and the collected data is given in Table 2.13.

Then we place these data (Table 2.13) in a data frame we called Data. Since the data are in different scales, each column is standardized using the function scale in R, and the first six observations of the scaled variables are given below. The complete code that provides the output given below is available in Appendix 2.

```
> Data=read.csv("Simulated_PCA.csv", header = T)
>
> ####We scale each column of the predictors
> Rscaled=scale(Data[,-1])
> head(Rscaled)
```

Table 2.13 Five independent variables in different scales

Line	Yield	PlantHeight	DaysFlowering	DaysMaturity	WeightFreshPlant
L1	5.49	179.98	64.6	119.39	43.86
L2	6.84	181.1	64.68	121.67	44.72
L3	6.75	181	64.38	120.1	45.36
L4	4.98	180.41	64.03	120.47	44.21
L5	8.36	180.89	66.13	122.3	45.42
L6	4.43	179.74	63.26	119.88	43.65
L7	6.67	180.49	64.83	120.22	44
L8	4.44	177.94	62.31	118.46	42.6
L9	5.62	178.91	63.41	120.64	44.07
L10	6.96	179.93	64.03	119.99	43.53
L11	5.91	181.21	64.03	120.86	43.82
L12	5.59	180.32	64.89	120.98	45.21
L13	5.27	180.97	63.75	120.14	44.29
L14	4.32	177.79	62.47	118.72	42.28
L15	5.48	180.04	63.82	120.08	43.86

```
          Yield PlantHeight DaysFlowering DaysMaturity WeightFreshPlant
[1,] -0.2814557  -0.06307151   0.57645346  -0.8738660    -0.2214903
[2,]  0.9159136   0.97575340   0.65900049   1.4162657     0.7373101
[3,]  0.8360890   0.88300118   0.34944911  -0.1607110     1.4508360
[4,] -0.7337952   0.33576305  -0.01169416   0.2109332     0.1687191
[5,]  2.2640628   0.78097373   2.15516550   2.0490652     1.5177291
[6,] -1.2216124  -0.28567685  -0.80620937  -0.3816886    -0.4556160
```

We then calculate the scaled variance (correlation) $\boldsymbol{Q} = \frac{1}{n-1}\boldsymbol{X}^{\mathrm{T}}\boldsymbol{X}$ and from this matrix we calculate the eigenvalue decomposition using the function eigen() of R. Next, we extract the eigenvectors and eigenvalues, and using the eigenvalue information, we calculate the variance of each principal component as shown below.

```
> n=nrow(Rscaled)
> Q_scaled=(t(Rscaled)%*%Rscaled)/(n-1) #Q_scaled=var(Rscaled)
> ####Eigenvalue decomposition
> SVD_Rscaled=eigen(Q_scaled)
>
> ####Extracting eigenvectors and eigenvalues
> EVectors=SVD_Rscaled$vectors
> Eigenvalues=SVD_Rscaled$values
> Standar.deviations.PC=sqrt(Eigenvalues)
> Standar.deviations.PC
[1] 2.0090648 0.6469991 0.4964878 0.4356803 0.3297472
```

Afterward, we manually calculate the principal components using the expression $\mathrm{PC} = \boldsymbol{XW}$, where $\boldsymbol{W}$ is the matrix of eigenvectors extracted in the previous code and denoted as EVectors. The calculation used here was the scaled matrix (Rscaled) instead of the original matrix of independent variables.

```
> ####Principal components for all the p=5 variables
> PCM=Rscaled%*%EVectors
> head(PCM)
            [,1]        [,2]       [,3]        [,4]        [,5]
[1,] -0.371645755  0.02096745  0.6294630  0.59140700 -0.58621466
[2,]  2.096405335 -0.02185667 -0.2605723 -0.54289807  0.12914892
[3,]  1.489539155 -0.30245825  0.5022823  0.73805410  0.79018009
[4,] -0.001842872 -0.78507631 -0.2113135 -0.06274286 -0.24361792
[5,]  3.931856138  1.09587334 -0.4321212  0.05736360 -0.17988334
[6,] -1.403180027 -0.72567496 -0.2400658 -0.12744453 -0.08832723
```

We then built the scree plot, which is one of the three tools used to select the number of principal components to maintain.

```
> ####Scree plot
> Ordered_Eigenvalues=sort(Eigenvalues,decreasing =T )
> plot(Ordered_Eigenvalues, type = "l",ylab="Variances",
xlab="Principal components")
```

Figure 2.2 shows that after two principal components, there are no significant gains in variance explained by adding more principal components. Due to this, we can select the first two principal components that explain 89.09% of the total variance of the complete data set.

Finally, the next part of the code selects only the first two principal components that will replace the whole matrix of scaled independent variables denoted here as Rscaled. Therefore, the selection process only consists of extracting the first two columns of the matrix that contain all the principal components, as shown in the following code. This reduced matrix is then replaced by the original matrix of independent variables as input in any statistical machine learning algorithm.

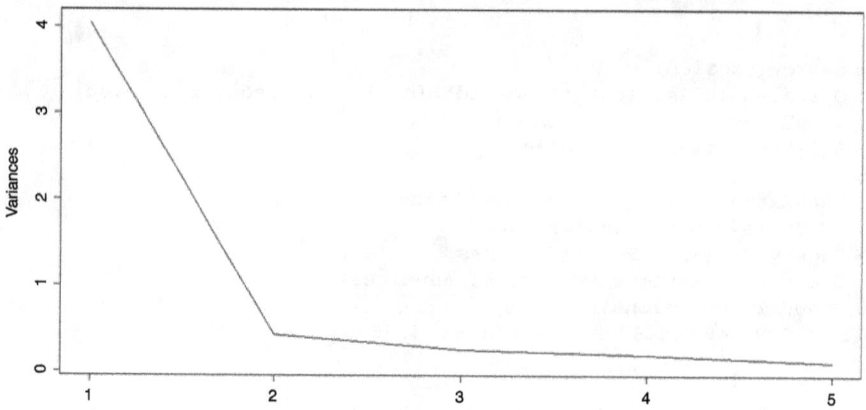

Fig. 2.2 Scree plot for the five independent variables in different scales

```
> ####Principal components for the first k=2 principal components
> X_star=PCM[,1:2]
> head(X_star)
          [,1]         [,2]
[1,]  -0.371645755  0.02096745
[2,]   2.096405335 -0.02185667
[3,]   1.489539155 -0.30245825
[4,]  -0.001842872 -0.78507631
[5,]   3.931856138  1.09587334
[6,]  -1.403180027 -0.72567496
```

It is important to point out that the code given above was used to manually calculate the new variables, called principal components. However, there are also many libraries of R that can be used to carry out this process automatically. One of these functions is the prcomp(), which performs a principal component analysis when the only input provided is the original scaled (or not scaled) matrix of original inputs. Next, we show how to use this function to perform a principal component analysis. Then, with the function summary(), the standard deviation of each of the five principal components is extracted, as well as the proportion of variance that explains each principal component and the cumulative proportion of variance explained for the principal component. Here, we can see that the standard deviations obtained using this function are the same as those obtained above when the principal components were extracted manually using the eigen() function.

```
> ####PCA analysis using the function prcomp
> PCA=prcomp(Rscaled)
> summary(PCA)
Importance of components:
                          PC1     PC2     PC3     PC4     PC5
Standard deviation     2.0091 0.64700 0.4965 0.43568 0.32975
Proportion of Variance 0.8073 0.08372 0.0493 0.03796 0.02175
Cumulative Proportion  0.8073 0.89099 0.9403 0.97825 1.00000
```

We then show how to extract all the principal components resulting from using the prcomp() function. The principal components extracted are clearly the same, except with negative values, which is not a problem since this does not alter the solution.

```
> ###Extracting the principal components####
> PC_All=PCA$x
> head(PCA$x)
          PC1          PC2         PC3         PC4          PC5
[1,] -0.371645755 -0.02096745  0.6294630 -0.59140700 -0.58621466
[2,]  2.096405335  0.02185667 -0.2605723  0.54289807  0.12914892
[3,]  1.489539155  0.30245825  0.5022823 -0.73805410  0.79018009
[4,] -0.001842872  0.78507631 -0.2113135  0.06274286 -0.24361792
[5,]  3.931856138 -1.09587334 -0.4321212 -0.05736360 -0.17988334
[6,] -1.403180027  0.72567496 -0.2400658  0.12744453 -0.08832723
```

Finally, we make the scree plot using the output of the prcomp() function, and the output of this figure is exactly the same as that given in Fig. 2.2.

```
> #####Variances of each principal component and scree plot####
> Var_PC=PCA$sdev*PCA$sdev
> plot(Var_PC, type = "l",ylab="Variances", xlab="Principal
components")
```

Appendix 1

```
rm()
library(synbreed)
####Loading the marker information
snp7 <- read.csv("MarkersToy.csv",header=T)
snp7=snp7[,-1]
snp7

####Set names for individuals
rownames(snp7) <- paste("ID",1:8,sep="")
pos.NA=which(snp7=="?_?", arr.ind=TRUE)
snp7[pos.NA]=NA
snp7
####Creating an object of class 'gpData'
gp <- create.gpData(geno=snp7)

####Recoding to 0, 1, and 2 values the genotypic data
gp.coded <-codeGeno(gp)
Geno_Recoded=gp.coded$geno
Geno_Recoded

####Recoding to values of 0, 1, and 2 the genotypic data and inputting
Imputed_Geno<-codeGeno(gp,impute=T,impute.type="random")
Imputed_Geno$geno
```

Appendix 2

```
rm(list=ls())
Data=read.csv("Simulated_PCA.csv", header = T)

####We scale each column of the predictors
Rscaled=scale(Data[,-1])
head(Rscaled)
n=nrow(Rscaled)
Q_scaled=(t(Rscaled)%*%Rscaled)/(n-1) #Q_scaled=var(Rscaled)
####Eigenvalue decomposition
SVD_Rscaled=eigen(Q_scaled)
```

```
####Extracting eigenvectors and eigenvalues
EVectors=SVD_Rscaled$vectors
Eigenvalues=SVD_Rscaled$values
Standar.deviations.PC=sqrt(Eigenvalues)
Standar.deviations.PC

####Principal components for all the p=5 variables
PCM=Rscaled%*%EVectors
head(PCM)

####Scree plot
Ordered_Eigenvalues=sort(Eigenvalues,decreasing =T )
plot(Ordered_Eigenvalues, type = "l",ylab="Variances",
xlab="Principal components")

####Principal components for the first k=2 principal components
X_star=PCM[,1:2]
head(X_star)

####PCA analysis using the function prcomp
PCA=prcomp(Rscaled)
summary(PCA)

###Extracting the principal components####
PC_All=PCA$x
head(PCA$x)

#####Variances of each principal component and scree plot####
Var_PC=PCA$sdev*PCA$sdev
plot(Var_PC, type = "l",ylab="Variances", xlab="Principal
components")
```

References

Austin PC, Brunner LJ (2004) Inflation of the type I error rate when a continuous confounding variable is categorized in logistic regression analyses. Stat Med 23(7):1159–1178

Griffiths JF, Griffiths AJ, Wessler SR, Lewontin RC, Gelbart WM, Suzuki DT, Miller JH (2005) An introduction to genetic analysis. Macmillan, New York

Henderson CR (1950) Estimation of genetic parameters. Ann Math Stat 21:309–310

Henderson CR (1963) Selection index and expected genetic advance. In: Hanson WD, Robinson HF (eds) Statistical genetics and plant breeding, Publication 982. Washington, DC, National Academy of Sciences, National Research Council, pp 141–163

Henderson CR (1973) Sire evaluation and genetic trends. In: Proceedings of the animal breeding and genetics symposium in honor of J. L. Lush. Blackburgh, Champaign, IL, American Society for Animal Science, pp 10–41

Henderson CR (1975) Best linear unbiased estimation and prediction under a selection model. Biometrics 31:423–447

Henderson CR (1984) Applications of linear models in animal breeding. University of Guelph, Guelph

Kuhn M, Johnson K (2013) Applied predictive modeling. Springer, New York
Littell RC, Milliken GA, Stroup WW, Wolfinger RD (1996) SAS for mixed models. SAS Institute, Inc., Cary, NC
Littell RC, Milliken GA, Stroup WW, Schabenberger O, Wolfinger RD (2006) SAS for mixed models, 2nd edn. SAS Institute, Cary, NC
Marsjan PA, Oldenbroek JK (2007) Molecular markers, a tool for exploring genetic diversity. In: The state of the world's animal genetic resources for food and agriculture. FAO, Rome. ISBN 9789251057629
Piepho HP, Möhring J, Melchinger AE, Büchse A (2008) BLUP for phenotypic selection in plant breeding and variety testing. Euphytica 161:209–228. https://doi.org/10.1007/s10681-007-9449-8
Schork NJ, Fallin D, Lanchbury S (2000) Single nucleotide polymorphisms and the future of genetic epidemiology. Clin Genet 58:250–264
Searle SR, Casella G, McCulloch CE (2006) Variance components. Wiley, Hoboken, NJ
Stonecking M (2001) From the evolutionary past. Nature 409:821–822
Stroup WW (2012) Generalized linear mixed models: modern concepts, methods and applications. CRC Press, Boca Raton, FL
The International SNP Map Working Group (2001) A map of human genome sequence variation containing 1.42 million single nucleotide polymorphisms. Nature 409:928–933
VanRaden PM (2008) Efficient methods to compute genomic predictions. J Dairy Sci 91:4414–4423
Zeger SL, Liang KY, Albert PS (1998) Models for longitudinal data: a generalized estimating equation approach. Biometrics 44(4):1049–1060

3

Fitting a Linear Regression Model: Major Concepts and Approaches

3.1 Definition of a Linear Multiple Regression Model

A linear multiple regression model (LMRM) is a useful tool for investigating linear relationships between two or more explanatory variables (inputs, features in machine learning literature) (X) and the conditional expected value of a response $E(Y/X)$. Due to its simplicity, adequate fitting, and easily interpretable results, this has been one of the most popular techniques for studying the association between variables. Specifically, regarding the latter task, this is a useful approach and an ideal (natural) starting point for studying more advanced methods (James et al. 2013) of association and prediction.

In this chapter, we review the main concepts and approaches for fitting a linear regression model.

3.2 Fitting a Linear Multiple Regression Model via the Ordinary Least Square (OLS) Method

In a general context, we have a covariate vector $X = (X_1, \ldots, X_p)^{\mathrm{T}}$ and we want to use this information to predict or explain how this variable affects a real-value response Y. The linear multiple regression model assumes a relationship given by

$$Y = \beta_0 + \sum_{j=1}^{p} X_j \beta_j + \epsilon, \tag{3.1}$$

where ϵ is a random error with mean 0, $E(\epsilon) = 0$ and is independent of X. This error is included in the model to capture measurement errors and the effects of other unregistered explanatory variables that can help to explain the mean response.

Then, the conditional mean of this model is $E(Y|X) = \beta_0 + \sum_{j=1}^{p} X_j\beta_j$ and the conditional distribution of Y given X is only affected by the information of X.

For estimating the parameters $\boldsymbol{\beta} = (\beta_0, \beta_1, \ldots, \beta_p)^{\mathrm{T}}$, usually we have a set of data $(\boldsymbol{x}_i^{\mathrm{T}}, y_i)$, $i = 1, \ldots, n$, often known as training data, where $\boldsymbol{x}_i = (x_{i1}, \ldots, x_{ip})^{\mathrm{T}}$ is a vector of features measurement and y_i is the response measurement corresponding to the ith individual drawn. The most common method for estimating $\boldsymbol{\beta}$ is the least squares method (OLS) that consists of taking the $\boldsymbol{\beta}$ value that minimizes the residual sum of squares defined as

$$\mathrm{RSS}(\boldsymbol{\beta}) = \sum_{i=1}^{n} (y_i - \beta_0 - \boldsymbol{x}_i^{\mathrm{T}}\boldsymbol{\beta}_0)^2 = (\boldsymbol{y} - \boldsymbol{X\beta})^{\mathrm{T}}(\boldsymbol{y} - \boldsymbol{X\beta}),$$

where $\boldsymbol{\beta}_0 = (\beta_1, \ldots, \beta_p)^{\mathrm{T}}$, $\boldsymbol{y} = (y_1, \ldots, y_n)^{\mathrm{T}}$ is the vector with the response values of all individuals, and $\boldsymbol{X}$ is an $n \times (p + 1)$ matrix that contains the information of the measured features of all individuals, including the intercept in the first entry:

$$\boldsymbol{X} = \begin{bmatrix} 1 & x_{11} & \cdots & x_{1p} \\ \vdots & \vdots & \vdots & \vdots \\ 1 & x_{n1} & \cdots & x_{np} \end{bmatrix}.$$

If the $\boldsymbol{X}$ matrix has full column rank, then by differentiating the residual sum of squares with respect to the $\boldsymbol{\beta}$ coefficients, we can find the set of $\boldsymbol{\beta}$ parameters that minimize the RSS($\boldsymbol{\beta}$),

$$\frac{\mathrm{RSS}(\boldsymbol{\beta})}{\partial\boldsymbol{\beta}} = \frac{(\boldsymbol{y} - \boldsymbol{X\beta})^{\mathrm{T}}(\boldsymbol{y} - \boldsymbol{X\beta})}{\partial\boldsymbol{\beta}} = \frac{\boldsymbol{y}^{\mathrm{T}}\boldsymbol{y} - 2\boldsymbol{y}^{\mathrm{T}}\boldsymbol{X\beta} + \boldsymbol{\beta}^{\mathrm{T}}(\boldsymbol{X}^{\mathrm{T}}\boldsymbol{X})\boldsymbol{\beta}}{\partial\boldsymbol{\beta}} = 2\left[(\boldsymbol{X}^{\mathrm{T}}\boldsymbol{X})\boldsymbol{\beta} - \boldsymbol{X}^{\mathrm{T}}\boldsymbol{Y}\right]$$

This derivative is also known as the gradient of the residual sum of squares. Then by setting the gradient of the residual sum of squares to zero, we obtain the normal equations

$$(\boldsymbol{X}^{\mathrm{T}}\boldsymbol{X})\boldsymbol{\beta} = \boldsymbol{X}^{\mathrm{T}}\boldsymbol{Y}$$

The solution to the normal equations is unique and gives the OLS estimator of $\boldsymbol{\beta}$

$$\widehat{\boldsymbol{\beta}} = (\boldsymbol{X}^{\mathrm{T}}\boldsymbol{X})^{-1}\boldsymbol{X}^{\mathrm{T}}\boldsymbol{y},$$

where super index -1 indicates the inversion matrix.

From the above assumptions, we can show that this estimator is unbiased

$$\begin{aligned} E\left(\widehat{\boldsymbol{\beta}}\right) &= E\left[(\boldsymbol{X}^{\mathrm{T}}\boldsymbol{X})^{-1}\boldsymbol{X}^{\mathrm{T}}\boldsymbol{y}\right] = E\left[(\boldsymbol{X}^{\mathrm{T}}\boldsymbol{X})^{-1}\boldsymbol{X}^{\mathrm{T}}(\boldsymbol{X\beta} + \epsilon)\right] \\ &= E\left[(\boldsymbol{X}^{\mathrm{T}}\boldsymbol{X})^{-1}\boldsymbol{X}^{\mathrm{T}}\boldsymbol{X\beta}\right] + (\boldsymbol{X}^{\mathrm{T}}\boldsymbol{X})^{-1}\boldsymbol{X}^{\mathrm{T}}E(\epsilon) = \boldsymbol{\beta}. \end{aligned}$$

and with the additional assumption that the observation responses y_i's are uncorrelated and have the same variance, $\mathrm{Var}(y_i) = \sigma^2$, we can also show that the variance–covariance matrix of this is

$$\mathrm{Var}\left(\widehat{\boldsymbol{\beta}}\right) = \sigma^2 \left(\boldsymbol{X}^{\mathrm{T}}\boldsymbol{X}\right)^{-1}.$$

When the input features only contain the information of a variable ($p = 1$), the resulting model is known as simple linear regression and can be easily visualized in the Cartesian plane. When $p = 2$, the above multiple linear regression describes a plane in the three-dimensional space (x_1, x_2, y). In general, the conditional expected value of this model defines a hyperplane in the p-dimensional space of the input variables (Montgomery et al. 2012).

The fitted values corresponding to all the training individuals are

$$\widehat{\boldsymbol{y}} = \boldsymbol{X}\widehat{\boldsymbol{\beta}} = \boldsymbol{X}\left(\boldsymbol{X}^{\mathrm{T}}\boldsymbol{X}\right)^{-1}\boldsymbol{X}^{\mathrm{T}}\boldsymbol{y} = \boldsymbol{H}\boldsymbol{y},$$

where the matrix $\boldsymbol{H} = \boldsymbol{X}(\boldsymbol{X}^{\mathrm{T}}\boldsymbol{X})^{-1}\boldsymbol{X}^{\mathrm{T}}$ is commonly called the hat matrix. This is because the vector of the observed response values is mapped by this expression to a vector of fitted values (Montgomery et al. 2012), in this way, puts the hat on $\boldsymbol{y}$ (Hastie et al. 2009). In a similar way, a predicted value of an arbitrary individual with feature $\boldsymbol{x}$ can be obtained by

$$\widehat{y}^* = \boldsymbol{x}^{*\mathrm{T}}\widehat{\boldsymbol{\beta}},$$

where $\boldsymbol{x}^* = (1, \boldsymbol{x}^{\mathrm{T}})^{\mathrm{T}}$.

An unbiased estimator for the common residual variance σ^2 is obtained by

$$\begin{aligned}
\widehat{\sigma}^2 &= \frac{1}{n-p-1}\sum_{i=1}^{n}(y_i - \widehat{y}_i)^2 \\
&= \frac{1}{n-p-1}\sum_{i=1}^{n} e_i^2 \\
&= \frac{1}{n-p-1}\boldsymbol{e}^{\mathrm{T}}\boldsymbol{e} \\
&= \frac{1}{n-p-1}(\boldsymbol{y} - \widehat{\boldsymbol{y}})^{\mathrm{T}}(\boldsymbol{y} - \widehat{\boldsymbol{y}}) \\
&= \frac{1}{n-p-1}\boldsymbol{y}^{\mathrm{T}}(\boldsymbol{I}_n - \boldsymbol{H})\boldsymbol{y},
\end{aligned}$$

where $e_i = y_i - \widehat{y}_i$ is known as the residual of the model corresponding to the individual i, $\boldsymbol{e} = \boldsymbol{y} - \widehat{\boldsymbol{y}}$ is the vector of all residual values, and $\boldsymbol{I}_n$ is the identity matrix of order $n \times n$.

The traditional inferential and prediction analysis for this model assumes that the random error ϵ is normally distributed with mean zero and variance σ^2. With this we can show that the OLS of beta coefficients, $\widehat{\boldsymbol{\beta}}$, is a random vector distributed according to a multivariate normal distribution with vector mean $\boldsymbol{\beta}$ and a variance–covariance matrix, as previously defined (Montgomery et al. 2012; Hastie et al. 2009; Rencher and Schaalje 2008). Another important fact that will be described in more detail in the next section, is that under the Gaussian assumption over errors, the OLS of $\boldsymbol{\beta}$ coincides with the maximum likelihood estimator.

We can also show that $(n-p-1)\widehat{\sigma}^2/\sigma^2$ is independent of $\widehat{\boldsymbol{\beta}}$ and distributed according to a Chi-squared distribution with $n-p-1$ degrees of freedom. Based on this and on the properties of the normal and t-student distributions, we show that for each $j=0,\ldots,p$, $T_j=\left(\widehat{\beta}_j-\beta_j\right)/\sqrt{c_{jj}\widehat{\sigma}^2}$, where c_{jj} is the $(j+1, j+1)$ elements of the matrix $(\boldsymbol{X}^{\mathrm{T}}\boldsymbol{X})^{-1}$, are random variables with a t-student distribution with $n-p-1$ degrees of freedom (t_{n-p-1}). That is, $T_j \sim t_{n-p-1}$ and $\sim$ stands for distributed as. From here, a $100(1-\alpha)\%$ confidence interval for a particular beta coefficient, β_j, is given by

$$\widehat{\beta}_j \pm t_{1-\alpha/2,\ n-p-1}\sqrt{c_{jj}\widehat{\sigma}^2},$$

where $t_{\alpha,\,n-p-1}$ is the α quantile of the t-student distribution with $n-p-1$ degrees of freedom. Similarly, a $100(1-\alpha)\%$ joint confidence region for all the beta coefficients, $\boldsymbol{\beta}$, is given if these values satisfy

$$\frac{\left(\widehat{\boldsymbol{\beta}}-\boldsymbol{\beta}\right)^{\mathrm{T}}\boldsymbol{X}^{\mathrm{T}}\boldsymbol{X}\left(\widehat{\boldsymbol{\beta}}-\boldsymbol{\beta}\right)}{(p+1)\widehat{\sigma}^2} \leq F^{p+1}_{1-\alpha,\ n-p-1},$$

where $F^{p+1}_{\alpha,n-p-1}$ denotes the α quantile of the F distribution with $p+1$ and $n-p-1$ degrees of freedom in the numerator and denominator, respectively (Rencher and Schaalje 2008).

In a similar way, to test a hypothesis over a specific beta coefficient, $H_{0j}=\beta_j=\beta_{j0}$, the following rule can be used: reject H_{0j} if $T_{j0}=\left(\widehat{\beta}_j-\beta_{j0}\right)/\sqrt{c_{jj}\widehat{\sigma}^2}$ is "large" in magnitude, that is, if $|T_{j0}|>t_{1-\alpha/2,\,n-p-1}$, where α is the desired level test. More generally, the test $H_0=\boldsymbol{W\beta}=\boldsymbol{w}$, where $\boldsymbol{W}$ is a $q\times(p+1)$ matrix of rank $q\leq p+1$, can be performed using the following rule:

$$\text{reject } H_0 \text{ if } F=\frac{n-p-1}{q}\frac{(\boldsymbol{W\beta}-\boldsymbol{w})^{\mathrm{T}}\left[\boldsymbol{W}\left(\boldsymbol{X}^{\mathrm{T}}\boldsymbol{X}\right)^{-1}\boldsymbol{W}^{\mathrm{T}}\right]^{-1}(\boldsymbol{W\beta}-\boldsymbol{w})}{\widehat{\sigma}^2} \geq F^{p+1}_{1-\alpha,n-p-1}.$$

3.3 Fitting the Linear Multiple Regression Model via the Maximum Likelihood (ML) Method

The maximum likelihood (ML) estimation is a more general and popular method for estimating the parameters of a model (Casella and Berger 2002). It consists of finding the parameter value that maximizes the "probability" of observed values in the sample under the adopted model. Specifically, if $\left(\boldsymbol{x}_i^{\mathrm{T}}, y_i\right)$, $i = 1, \ldots, n$, is a set of observations from a multiple linear regression model (3.1) with homoscedastic and uncorrelated errors, the MLE of $\boldsymbol{\beta}$ and σ^2, $\widehat{\boldsymbol{\beta}}$ and $\widehat{\sigma}^2$, of this model is defined as

$$\left(\widehat{\boldsymbol{\beta}}^{\mathrm{T}}, \widehat{\sigma}^2\right) = \arg \max_{\boldsymbol{\beta}, \sigma^2} L\left(\boldsymbol{\beta}, \sigma^2; \boldsymbol{y}, \boldsymbol{X}\right),$$

where $L(\boldsymbol{\beta}, \sigma^2; \boldsymbol{y}, \boldsymbol{X})$ is the likelihood function of the parameters, which is the probability of the observed response values but viewed as a function of the parameters

$$L\left(\boldsymbol{\beta}, \sigma^2; \boldsymbol{y}, \boldsymbol{X}\right) = \left(\frac{1}{\sqrt{2\pi\sigma^2}}\right)^n \exp\left[-\frac{1}{2\sigma^2}(\boldsymbol{y} - \boldsymbol{X\beta})^{\mathrm{T}}(\boldsymbol{y} - \boldsymbol{X\beta})\right].$$

Then, the $\log(L(\boldsymbol{\beta}, \sigma^2; \boldsymbol{y}, \boldsymbol{X}))$ is equal to

$$\log\left(L\left(\boldsymbol{\beta}, \sigma^2; \boldsymbol{y}, \boldsymbol{X}\right)\right) = -\frac{n}{2}\log(2\pi) - n\log(\sigma) - \frac{1}{2\sigma^2}(\boldsymbol{y} - \boldsymbol{X\beta})^{\mathrm{T}}(\boldsymbol{y} - \boldsymbol{X\beta})$$

To find the maximum of σ^2 and $\boldsymbol{\beta}$, we get the derivative of $\log\left(L\left(\widehat{\boldsymbol{\beta}}, \sigma^2; \boldsymbol{y}, \boldsymbol{X}\right)\right)$ with regard to these parameters

$$\frac{\log\left(L\left(\boldsymbol{\beta}, \sigma^2; \boldsymbol{y}, \boldsymbol{X}\right)\right)}{\partial \boldsymbol{\beta}} = \frac{\left[\left(\boldsymbol{X}^{\mathrm{T}}\boldsymbol{X}\right)\boldsymbol{\beta} - \boldsymbol{X}^{\mathrm{T}}\boldsymbol{Y}\right]}{\sigma^2}$$

$$\frac{\log\left(L\left(\boldsymbol{\beta}, \sigma^2; \boldsymbol{y}, \boldsymbol{X}\right)\right)}{\partial \sigma^2} = -\frac{n}{2\sigma^2} + \frac{1}{2\sigma^4}(\boldsymbol{y} - \boldsymbol{X\beta})^{\mathrm{T}}(\boldsymbol{y} - \boldsymbol{X\beta})$$

Now, by setting these derivatives equal to zero and solving the resulting equations for $\boldsymbol{\beta}$ and σ^2, we found that the estimates of these parameters are

$$\widehat{\boldsymbol{\beta}} = \left(\boldsymbol{X}^{\mathrm{T}}\boldsymbol{X}\right)^{-1}\boldsymbol{X}^{\mathrm{T}}\boldsymbol{y}$$

$$\widehat{\sigma}^2 = \frac{1}{n}\left(\boldsymbol{y} - \boldsymbol{X}\widehat{\boldsymbol{\beta}}\right)^{\mathrm{T}}\left(\boldsymbol{y} - \boldsymbol{X}\widehat{\boldsymbol{\beta}}\right).$$

From this we can see that for each value of σ^2, the value of $\boldsymbol{\beta}$ that maximizes the likelihood is the same value that maximizes $-\frac{1}{2\sigma^2}(\boldsymbol{y} - \boldsymbol{X\beta})^{\mathrm{T}}(\boldsymbol{y} - \boldsymbol{X\beta})$, which in turn minimizes $(\boldsymbol{y} - \boldsymbol{X\beta})^{\mathrm{T}}(\boldsymbol{y} - \boldsymbol{X\beta})$, which is precisely the OLS of $\boldsymbol{\beta}, \widehat{\boldsymbol{\beta}}$. But when equating the derivative of $\log\left(L\left(\widehat{\boldsymbol{\beta}}, \sigma^2; \boldsymbol{y}, \boldsymbol{X}\right)\right)$ to zero and solving for σ^2, the value of σ^2 that maximizes $L\left(\widehat{\boldsymbol{\beta}}, \sigma^2; \boldsymbol{y}, \boldsymbol{X}\right)$ is $\widehat{\sigma}^2 = \frac{1}{n}\left(\boldsymbol{y} - \boldsymbol{X}\widehat{\boldsymbol{\beta}}\right)^{\mathrm{T}}\left(\boldsymbol{y} - \boldsymbol{X}\widehat{\boldsymbol{\beta}}\right)$.

Finally,

$$L(\boldsymbol{\beta}, \sigma^2; \boldsymbol{y}, \boldsymbol{X}) \leq L\left(\widehat{\boldsymbol{\beta}}, \sigma^2; \boldsymbol{y}, \boldsymbol{X}\right) \leq L\left(\widehat{\boldsymbol{\beta}}, \widehat{\sigma}^2; \boldsymbol{y}, \boldsymbol{X}\right)$$

and from here, the MLE of $\boldsymbol{\beta}$ and σ^2 are $\widehat{\boldsymbol{\beta}}$ and $\widehat{\sigma}^2$, because it can be shown that the values of parameters that maximize the likelihood are unique when the design matrix $\boldsymbol{X}$ is of full column rank.

3.4 Fitting the Linear Multiple Regression Model via the Gradient Descent (GD) Method

The steepest descent method, also known as the gradient descent (GD) method, is a first-order iterative algorithm for minimizing a function (f). It is a central mechanism in statistical learning to training models (to estimate the parameters), for example, in neuronal networks and penalized regression models (Ridge and Lasso). It consists of successively updating the argument of the objective function in the direction of the steepest descent (along the negative of the gradient of the function), that is, in the direction in which f decreases most rapidly (Haykin 2009; Nocedal and Wright 2006). Specifically, each step of this algorithm is described by

$$\eta_{t+1} = \eta_t - \alpha \nabla f(\eta_t),$$

where $\nabla f(\eta_t)$ is the gradient vector of f evaluated in the current value η_t and α is a step size or learning rate parameter, which greatly determines the convergence behavior toward an optimal solution (Haykin 2009; Beysolow II 2017) and in neural networks it is popular for setting this at a small, fixed value (Warner and Misra 1996; Goodfellow et al. 2016). The learning rate parameter can be adaptative as well, that is, can be allowed to change at each step. For example, in the library Keras (see Chap. 11) that can be used for implementing and training neuronal networks models, there are several optimizers based on an adaptive gradient descendent algorithm such as Adam Adgrad, Adadelta, RMSprop, among others (Allaire and Chollet 2019). The ideal value of the step size would be the value that gives the larger reduction in each step, that is, the value of α that minimizes $f(\eta_t - \alpha \nabla f(\eta_t))$, which in general is difficult and expensive to obtain (Nocedal and Wright 2006).

Although the use of this algorithm could be avoided in an MLR, especially in small data sets, and also because of its slow convergence in linear systems (Burden and Faires 2011), here we will describe how this works when finding the optimal beta coefficients in this model. First, the gradient of the residual sum of squares is given by

$$\nabla \text{RSS}(\boldsymbol{\beta}) = 2\left(\boldsymbol{X}^{\text{T}}\boldsymbol{X}\boldsymbol{\beta} - \boldsymbol{X}^{\text{T}}\boldsymbol{y}\right).$$

Then, the next update of beta coefficients in the gradient descent algorithm in this model is given by

$$\begin{aligned}\boldsymbol{\beta}_{t+1} &= \boldsymbol{\beta}_t - 2\alpha\left(\boldsymbol{X}^{\text{T}}\boldsymbol{X}\boldsymbol{\beta}_t - \boldsymbol{X}^{\text{T}}\boldsymbol{y}\right)\\ &= \boldsymbol{\beta}_t - 2\alpha\boldsymbol{X}^{\text{T}}(\boldsymbol{X}\boldsymbol{\beta}_t - \boldsymbol{y})\\ &= \boldsymbol{\beta}_t + 2\alpha\boldsymbol{X}^{\text{T}}\boldsymbol{e}_t,\end{aligned}$$

where $\boldsymbol{e}_t = \boldsymbol{y} - \boldsymbol{X}\boldsymbol{\beta}_t$ is the vector of residuals that is obtained in the current iteration. One way to speed up the convergence of the algorithm is by choosing the ideal learning rate in each step, which, as was described before, is given by the value of α that minimizes $f(\eta_t - \alpha \nabla f(\eta_t))$, and in this case for the MLR model is given by (Nocedal and Wright 2006):

$$\alpha_t = \frac{\boldsymbol{e}_t^{\text{T}}\boldsymbol{X}\boldsymbol{X}^{\text{T}}\boldsymbol{e}_t}{\boldsymbol{e}_t^{\text{T}}\left(\boldsymbol{X}\boldsymbol{X}^{\text{T}}\right)^2\boldsymbol{e}_t}.$$

Example 1 For numerical illustration, we considered a synthetic data set that consists of 100 observations and two covariates. The scatter plots in Fig. 3.1 show how the response variable (y) is related to the two covariates (x_1, x_2). By setting a value of 10^{-2} for the learning rate parameter, and as the stopping criterion a tolerance of 10^{-8} for the maximum norm of the difference between the current and next vector value, the beta coefficient obtained with the GD method is $\widehat{\boldsymbol{\beta}} = (5.0460764, 0.8551383, 2.1903356)$. For these synthetic examples, 12 iterations were necessary, while by changing the learning rate parameter to 10^{-3}, the number of iterations increased to 185, but we practically got the same results. Now, by using the "optimal" learning rate parameter described before for MLR with the same tolerance error (10^{-8}), the number of required iterations up to convergence is reduced to only 10 iterations. In general, the performance of the gradient descent depends greatly on the objective function and can be affected by the characteristics of the model, the dispersion of the data (explained variance of the predictors), and the dependence between the predictors, among others.

In the data set used in this example, the covariates are independent and the proportion of explained variances by the predictor is about 79% of the total variance of the response. By changing to a pair of moderately correlated covariates with correlation 0.75, while holding the same beta coefficient values, the variance of the residual (1.44), and the same sample size, we generate data where a greater proportion of variance is explained by the covariates (85.6%), but when applying the

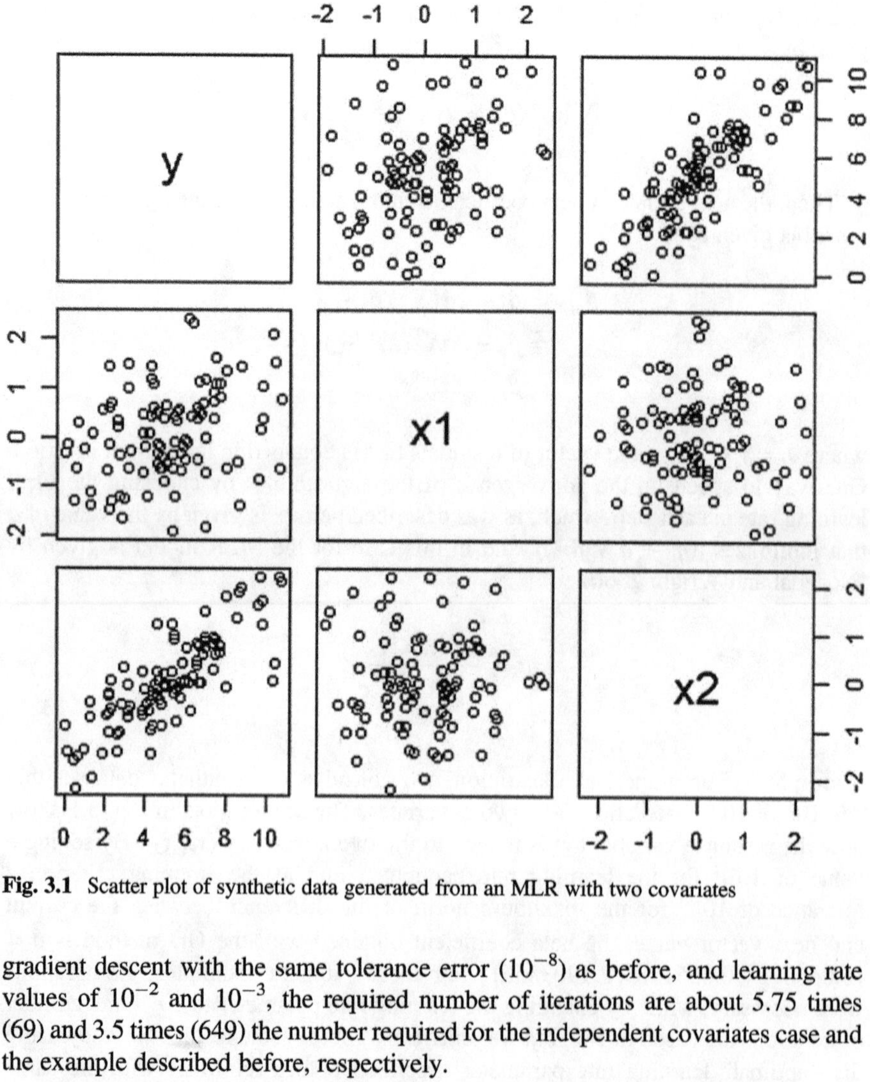

Fig. 3.1 Scatter plot of synthetic data generated from an MLR with two covariates

gradient descent with the same tolerance error (10^{-8}) as before, and learning rate values of 10^{-2} and 10^{-3}, the required number of iterations are about 5.75 times (69) and 3.5 times (649) the number required for the independent covariates case and the example described before, respectively.

Continuing with the last case of dependent variables, when using the optimal learning rate described before for the MLR, the number of iterations is reduced to 60, 9 less than when using the constant learning rate 10^{-2}.

By multiplying the beta coefficients used before by sqrt(0.1), the proportion of explained variance by the covariates is reduced to 27.30% and 37.2% in the same independent covariate (E3) and the same correlated covariate (E4) scenarios described before, respectively. With a tolerance error of 10^{-8} and with a learning rate equal to 10^{-2}, the required number of iterations are 183 and 66, for scenarios E3 and E4, respectively, while for a learning rate of 10^{-3} the required number of iterations are 617 and 1638. When using the "optimal" learning rate parameter, the required number of iterations is reduced to 17 and 56 for scenarios E3 and E4, respectively.

The R code used for implementing the GD method is given next.

```
#################R code for Example 1 ################################
rm(list=ls())
library(mvtnorm)
set.seed(1)
X = cbind(1,rmvnorm(100,c(0,0),diag(2)))
#Uncomment the next three lines code to simulate dependent covariables
#Sigma_X =0.75+0.25*diag(2)
#L = t(chol(Sigma_X))
#X = X%*%t(L)


betav = c(5,1,2.1)
#Uncomment the next line code to reduce the value of the beta cofficients
and reduce the proportion of variance of the response explained by the
features
betav = sqrt(0.1)*beta
y = X%*%betav + rnorm(100,0,1.2)
dat = data.frame(y=y,x1 = X[,2],x2=X[,3])
plot(dat)


alpha = 1e-2
#alpha = 1e-3
tol    = 1e-8
p = 2
betav_0 = c(mean(y),rep(0,p))
tol.e = 1
Iter = 0
tX = t(X)
XtX = X%*%t(X)
while(tol<tol.e)
{
 Iter = Iter + 1
 e = y-X%*%betav_0
 #Uncomment the next line code to use the optimal learning rate
 #alpha = (t(e)%*%XtX%*%e/(t(e)%*%(XtX)%*%XtX%*%e))[1,1]
 betav_t = betav_0 + alpha*tX%*%e
 tol.e = max(abs(betav_t-betav_0))
 betav_0 = betav_t
}
betav_t
tol.e
Iter
```

This code is only for illustrative purposes, that is, to illustrate in a very transparent way how the GD method can be implemented. Of course the existing statistical machine learning software programs implement this method and so there is no need to use this program for real applications, since the existing software programs that implement this method do a lot of work more efficiently and in a more user-friendly way.

3.5 Advantages and Disadvantages of Standard Linear Regression Models (OLS and MLR)

The MLR is a simple and computationally appealing class of models, but with many predictors (relative to the sample size) or nearly dependent features, it may result in large prediction error and/or large predictive intervals (Wakefield 2013). To appreciate the latter case (nearly dependent features), consider the spectral decomposition $\boldsymbol{X}^{\mathrm{T}}\boldsymbol{X} = \boldsymbol{\Gamma}\boldsymbol{\Lambda}\boldsymbol{\Gamma}^{\mathrm{T}}$, where $\boldsymbol{\Lambda} = \mathrm{Diag}(\lambda_0, \ldots, \lambda_p)$ is a diagonal matrix with the eigenvalues of $\boldsymbol{X}^{\mathrm{T}}\boldsymbol{X}$ in decreasing order and $\boldsymbol{\Gamma}$ is an orthogonal matrix with columns corresponding to eigenvectors of $\boldsymbol{X}^{\mathrm{T}}\boldsymbol{X}$. Then the obtained variance–covariance matrix of the OLS estimator of $\widehat{\boldsymbol{\beta}}$ can be expressed as

$$\mathrm{Var}\left(\widehat{\boldsymbol{\beta}}\right) = \sigma^2\left(\boldsymbol{\Gamma}\boldsymbol{\Lambda}\boldsymbol{\Gamma}^{\mathrm{T}}\right)^{-1} = \sigma^2\boldsymbol{\Gamma}\boldsymbol{\Lambda}^{-1}\boldsymbol{\Gamma}^{\mathrm{T}}.$$

When the features are nearly dependent, some λ_j's will be "close" to zero and consequently the variance of some $\widehat{\beta}_j^{\prime}$s will be high; this is even greater when the linear dependence of the features is strong (Wakefield 2013; Christensen 2011). This strong dependence between features is a problem of the OLS in MLR that is also reflected in the quality of the prediction performance, for example, when this is measured by the conditional expected prediction error (EPE) or mean squared error prediction that for an individual with feature $\boldsymbol{x}_o$ is given by

$$\begin{aligned}
\mathrm{EPE}(\boldsymbol{x}_o) &= E_{\boldsymbol{Y},Y_o|\boldsymbol{X},\boldsymbol{x}_o}\left[\left(Y_o - \boldsymbol{x}_o^{*\mathrm{T}}\widehat{\boldsymbol{\beta}}\right)^2\right] \\
&= E_{\boldsymbol{Y},Y_o|\boldsymbol{X},\boldsymbol{x}_o}\left\{\left[\left(Y_o - E(Y_0|\boldsymbol{x}_0) + E(Y_0|\boldsymbol{x}_0) - \boldsymbol{x}_o^{*\mathrm{T}}\widehat{\boldsymbol{\beta}}\right)\right]^2\right\} \\
&= E_{\boldsymbol{Y},Y_o|\boldsymbol{X},\boldsymbol{x}_o}\left[\left(Y_o - E\left(Y_0|\boldsymbol{x}_0^{*\mathrm{T}}\right)\right)^2\right] + 2E_{Y_o|\boldsymbol{x}_o}\left[\left(Y_o - E\left(Y_0|\boldsymbol{x}_0^{*\mathrm{T}}\right)\right)\right]\left[E\left(Y_0|\boldsymbol{x}_0^{*\mathrm{T}}\right) - E_{\boldsymbol{Y}|\boldsymbol{X}}\left(\boldsymbol{x}_o^{*\mathrm{T}}\widehat{\boldsymbol{\beta}}\right)\right] \\
&\quad + E_{\boldsymbol{Y},Y_o|\boldsymbol{X}}\left[\left(E\left(Y_0|\boldsymbol{x}_0^{*\mathrm{T}}\right) - \boldsymbol{x}_o^{*\mathrm{T}}\widehat{\boldsymbol{\beta}}\right)^2\right] \\
&= \sigma^2 + E_{\boldsymbol{Y},Y_o|\boldsymbol{X},\boldsymbol{x}_o}\left[\left(\boldsymbol{x}_o^{*\mathrm{T}}\boldsymbol{\beta} - \boldsymbol{x}_o^{*\mathrm{T}}\widehat{\boldsymbol{\beta}}\right)^2\right] = \sigma^2 + Var\left(\boldsymbol{x}_o^{*\mathrm{T}}\widehat{\boldsymbol{\beta}}|\boldsymbol{x}_o\right) \\
&= \sigma^2 + \sigma^2\boldsymbol{x}_o^{*\mathrm{T}}\boldsymbol{\Gamma}\boldsymbol{\Lambda}^{-1}\boldsymbol{\Gamma}^{\mathrm{T}}\boldsymbol{x}_o^{*} \\
&= \sigma^2\left(1 + \sum_{j=0}^{p}\frac{\left(x_{oj}^{**}\right)^2}{\lambda_j}\right),
\end{aligned}$$

where $\boldsymbol{x}_o^{**} = \boldsymbol{\Gamma}^{\mathrm{T}}\boldsymbol{x}_o^{*} = \left(x_{o0}^{**}, \ldots, x_{op}^{**}\right)^{\mathrm{T}}$. This means that the average loss incurred (squared difference between the value to be predicted and the predicted value) by predicting Y_0 with its estimated mean under the MLR, $\boldsymbol{x}_o^{*\mathrm{T}}\widehat{\boldsymbol{\beta}}$, is composed of intrinsic or irreducible data noise (first term) and the variance of $\boldsymbol{x}_o^{*\mathrm{T}}\widehat{\boldsymbol{\beta}}$ (second term). The former cannot be avoided no matter how well the mean value of $Y_0 \mid x_0$, $E(Y_0 | x_0)$, is estimated, and the latter increases as the dependence of features is stronger. From this, it is apparent that the EPE is also affected by the strong dependence between features, which is a problem of the OLS in an MLR in a prediction context.

3.6 Regularized Linear Multiple Regression Model

3.6.1 *Ridge Regression*

Ridge regression, originally proposed as a method to combat multicollinearity, is also a common approach for controlling overfitting in an MLR model (Christensen 2011). It translates the OLS problem into the minimization of the penalized residual sum of squares defined as

$$\mathrm{PRSS}_{\lambda}(\boldsymbol{\beta}) = \sum_{i=1}^{n}\left(y_i - \beta_0 - \sum_{j=1}^{p} x_{ij}\beta_j\right)^2 + \lambda\sum_{j=1}^{p}\beta_j^2,$$

where $\lambda \geq 0$ is known as the regularization or tuning parameter, which determines the level or degree to which the beta coefficients are shrunk toward zero. When $\lambda = 0$, the OLS is the solution to the beta coefficients, but when λ is large, the $\mathrm{PRSS}_{\lambda}(\boldsymbol{\beta})$ is dominated by the penalization term, and the OLS solution has to shrink toward 0 (Christensen 2011). In general, when the number of parameters to be estimated is larger than the number of observations, the estimator can be highly variable. In this situation, the intuition of Ridge regression tries to alleviate this by constraining the sum of squares for the beta coefficients.

Note that $\mathrm{PRSS}_{\lambda}(\boldsymbol{\beta})$ can be expressed as

$$\mathrm{PRSS}_{\lambda}(\boldsymbol{\beta}) = \mathrm{RSS}(\boldsymbol{\beta}) + \lambda\boldsymbol{\beta}^{\mathrm{T}}\boldsymbol{D}\boldsymbol{\beta},$$

where $\boldsymbol{D} =$ diag $(0, 1, \ldots, 1)$ is an identity matrix of dimension $(p + 1) \times (p + 1)$ but with one zero in its first entry. Then, the gradient of $\mathrm{RSS}_{\lambda}(\boldsymbol{\beta})$, that is, the first derivative with regard to $\boldsymbol{\beta}$ of $\mathrm{RSS}_{\lambda}(\boldsymbol{\beta})$, is

$$\nabla\mathrm{PRSS}_{\lambda}(\boldsymbol{\beta}) = 2\left(\boldsymbol{X}^{\mathrm{T}}\boldsymbol{X}\boldsymbol{\beta} - \boldsymbol{X}^{\mathrm{T}}\boldsymbol{y}\right) + 2\lambda\boldsymbol{D}\boldsymbol{\beta}.$$

Solving $\nabla\mathrm{PRSS}_{\lambda}(\boldsymbol{\beta}) = \mathbf{0}$, the Ridge solution is given by

$$\widehat{\boldsymbol{\beta}}^{R}(\lambda) = \underset{\boldsymbol{\beta}}{\mathrm{argmin}}\mathrm{PRSS}_{\lambda}(\boldsymbol{\beta}) = \left(\boldsymbol{X}^{\mathrm{T}}\boldsymbol{X} + \lambda\boldsymbol{D}\right)^{-1}\mathbf{X}^{\mathrm{T}}\boldsymbol{y}.$$

This is a biased estimator of $\boldsymbol{\beta}$ because the conditional expected value is given by

$$E\left[\widehat{\boldsymbol{\beta}}^{R}(\lambda)\right] = \left(\boldsymbol{X}^{\mathrm{T}}\boldsymbol{X} + \lambda\boldsymbol{D}\right)^{-1}\mathbf{X}^{\mathrm{T}}\boldsymbol{X}\boldsymbol{\beta}$$

but as will be described later, relative to the OLS estimator, by introducing a "small" bias, the variance or/and the EPE of this method could potentially be reduced (Wakefield 2013).

By using the method of Lagrange multipliers, the Ridge regression estimates of the $\boldsymbol{\beta}$ coefficients can be reformulated in a similar way to the OLS problem, but subject to the condition that the magnitude of the $\boldsymbol{\beta}_0 = (\beta_1, \ldots, \beta_p)^{\mathrm{T}}$ be less or equal to $t(\lambda)^{\frac{1}{2}}$, that is,

$$\widehat{\boldsymbol{\beta}}^{R}(\lambda) = \underset{\boldsymbol{\beta}}{\operatorname{argmin}}\mathrm{RSS}(\boldsymbol{\beta})$$

$$\text{subject to} \sum_{j=1}^{p}\beta_j^2 \le t(\lambda),$$

where $t(\lambda)$ is a one-to-one function that produces an equivalent definition to the penalized OLS presentation of the Ridge regression described before (Wakefield 2013; Hastie et al. 2009, 2015). This constrained reformulation gives a more transparent role than the one played by the tuning parameter, and among other things, suggests a convenient and common way of redefining the Ridge estimator by standardizing the variables when these are of very different scales.

A graphic representation of this constraint problem for $\boldsymbol{\beta}_0 = 0$ and $p = 2$ is given in Fig. 3.2, where the nested ellipsoids correspond to contour plots of $\mathrm{RSS}(\boldsymbol{\beta})$ and the green region is the restriction with $t(\lambda) = 3^2$, which contains the Ridge solution.

The MLR defined in (3.1) but now defined with the standardized variables is expressed as

$$\begin{aligned} \boldsymbol{y} &= \mathbf{1}_n\mu + \boldsymbol{X}_{1s}\boldsymbol{\beta}_{0s} + \epsilon \\ &= \boldsymbol{X}_s\boldsymbol{\beta}_s + \epsilon, \end{aligned}$$

where $\mathbf{1}_n$ is the column vector with 1's in all its entries, $\boldsymbol{X}_{1s} = \begin{bmatrix} x_{11s} & \cdots & x_{1ps} \\ \vdots & \vdots & \vdots \\ x_{n1s} & \cdots & x_{nps} \end{bmatrix}$, $x_{ijs} = \left(x_{ij} - \bar{x}_j\right)/s_j$, $s_j = \sqrt{\sum_{i=1}^{n}\left(x_{ij} - \bar{x}_j\right)^2/n}$, $j = 1, \ldots, p$; $\boldsymbol{X}_s = [\mathbf{1}_n\ \boldsymbol{X}_{1s}]$; $\boldsymbol{\beta}_s = \left(\mu,\ \boldsymbol{\beta}_{0s}^{\mathrm{T}}\right)^{\mathrm{T}}$; and $\boldsymbol{\beta}_{0s} = (\beta_{1s}, \ldots, \beta_{ps})^{\mathrm{T}}$.

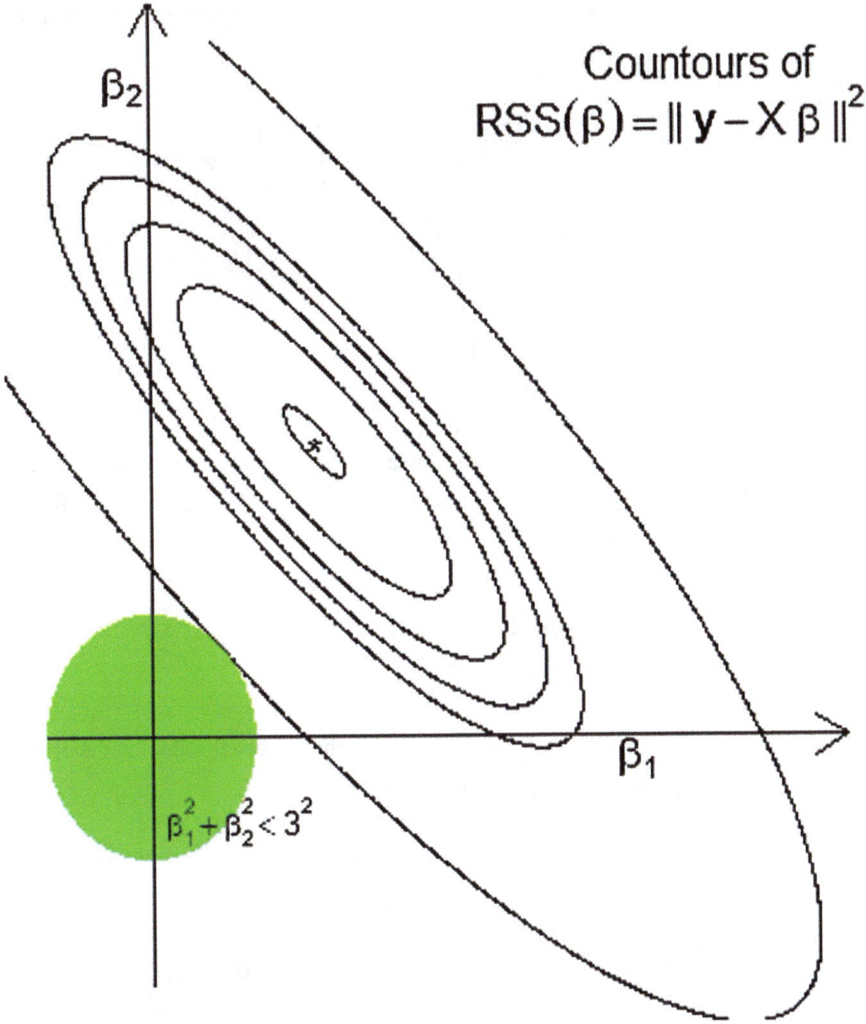

Fig. 3.2 Graphic representation of the Ridge solution of the OLS with restriction $\sum_{j=1}^{p}\beta_j^2 < 3^2$. The green region contains the Ridge solution for $t(\lambda) = 3^2$

Then, the redefined penalized residual sum squared under this model is

$$\begin{aligned}\mathrm{PRSS}_\lambda(\boldsymbol{\beta}_s) &= \sum_{i=1}^{n}\left(y_i - \mu - \sum_{j=1}^{p} x_{ijs}\beta_{js}\right)^2 + \lambda\sum_{j=1}^{p}\beta_{js}^2 \\ &= \left(\boldsymbol{y} - \boldsymbol{X}_s^{\mathrm{T}}\boldsymbol{\beta}_s\right)^{\mathrm{T}}\left(\boldsymbol{y} - \boldsymbol{X}_s^{\mathrm{T}}\boldsymbol{\beta}_s\right) + \lambda\boldsymbol{\beta}_{0s}{}^{\mathrm{T}}\boldsymbol{D}\boldsymbol{\beta}_{0s}.\end{aligned}$$

The Ridge solution under this redefinition is like the one given before, but now

$$
\begin{aligned}
\widehat{\boldsymbol{\beta}}_s^R(\lambda) &= (\boldsymbol{X}_s^{\mathrm{T}}\boldsymbol{X}_s + \lambda\boldsymbol{D})^{-1}\boldsymbol{X}_s^{\mathrm{T}}\boldsymbol{y} \\
&= \left(\begin{bmatrix} 1_n^{\mathrm{T}} \\ \boldsymbol{X}_{1s}^{\mathrm{T}} \end{bmatrix}[1_n\, \boldsymbol{X}_{1s}] + \lambda\boldsymbol{D}\right)^{-1}\begin{bmatrix} 1_n^{\mathrm{T}} \\ \boldsymbol{X}_{1s}^{\mathrm{T}} \end{bmatrix}\boldsymbol{y} \\
&= \begin{bmatrix} n & 0_n^{\mathrm{T}} \\ 0_n & \boldsymbol{X}_{1s}^{\mathrm{T}}\boldsymbol{X}_{1s} + \lambda\boldsymbol{I}_p \end{bmatrix}^{-1}\begin{bmatrix} 1_n^{\mathrm{T}}y \\ \boldsymbol{X}_{1s}^{\mathrm{T}}y \end{bmatrix} \\
&= \begin{bmatrix} \bar{y}_n \\ \widehat{\boldsymbol{\beta}}_{0s}(\lambda) \end{bmatrix},
\end{aligned}
$$

where $\bar{y}_n = \sum_{i=1}^{n} y_i/n$ is the sample mean of the responses and $\widehat{\boldsymbol{\beta}}_{0s}(\lambda) = (\boldsymbol{X}_{1s}^{\mathrm{T}}\boldsymbol{X}_{1s} + \lambda\boldsymbol{I}_p)^{-1}\boldsymbol{X}_{1s}^{\mathrm{T}}\boldsymbol{y}$ is the Ridge estimator of $\boldsymbol{\beta}_{0s}$. The mean value of this Ridge solution is

$$
\mathrm{E}\left[\widehat{\boldsymbol{\beta}}_s^R(\lambda)\right] = \begin{bmatrix} \mu \\ E\left[\widehat{\boldsymbol{\beta}}_{0s}(\lambda)\right] \end{bmatrix},
$$

where $E\left[\widehat{\boldsymbol{\beta}}_{0s}(\lambda)\right] = (\boldsymbol{X}_{1s}^{\mathrm{T}}\boldsymbol{X}_{1s} + \lambda\boldsymbol{I}_p)^{-1}\boldsymbol{X}_{1s}^{\mathrm{T}}\boldsymbol{X}_{1s}\boldsymbol{\beta}_{0s}$ is the expected value of the Ridge estimator of $\boldsymbol{\beta}_{0s}$. The variance–covariance matrix is

$$
\begin{aligned}
\mathrm{Var}\left(\widehat{\boldsymbol{\beta}}_s^R(\lambda)\right) &= (\boldsymbol{X}_s^{\mathrm{T}}\boldsymbol{X}_s + \lambda\boldsymbol{D})^{-1}\boldsymbol{X}_s^{\mathrm{T}}\mathrm{Var}(\boldsymbol{y})\boldsymbol{X}_s(\boldsymbol{X}_s^{\mathrm{T}}\boldsymbol{X}_s + \lambda\boldsymbol{D})^{-1} \\
&= \sigma^2\begin{bmatrix} n & 0_n^{\mathrm{T}} \\ 0_n & \boldsymbol{X}_{1s}^{\mathrm{T}}\boldsymbol{X}_{1s} + \lambda\boldsymbol{I}_p \end{bmatrix}^{-1}\begin{bmatrix} 1_n^{\mathrm{T}} \\ \boldsymbol{X}_{1s}^{\mathrm{T}} \end{bmatrix}[1_n\ \boldsymbol{X}_{1s}]\begin{bmatrix} n & 0_n^{\mathrm{T}} \\ 0_n & \boldsymbol{X}_{1s}^{\mathrm{T}}\boldsymbol{X}_{1s} + \lambda\boldsymbol{I}_p \end{bmatrix}^{-1} \\
&= \sigma^2\begin{bmatrix} 1/n & 0_n^{\mathrm{T}} \\ 0_n & \mathrm{Var}\left(\widehat{\boldsymbol{\beta}}_{0s}(\lambda)\right) \end{bmatrix},
\end{aligned}
$$

where $\mathrm{Var}\left(\widehat{\boldsymbol{\beta}}_{0s}(\lambda)\right) = (\boldsymbol{X}_{1s}^{\mathrm{T}}\boldsymbol{X}_{1s} + \lambda\boldsymbol{I}_p)^{-1}\boldsymbol{X}_{1s}^{\mathrm{T}}\boldsymbol{X}_{1s}(\boldsymbol{X}_{1s}^{\mathrm{T}}\boldsymbol{X}_{1s} + \lambda\boldsymbol{I}_p)^{-1}$. So, because in this standardized way, the Ridge solution of the intercept (μ) is the sample mean of the observed responses, and the correlation of this with the rest of the estimated parameters ($\widehat{\boldsymbol{\beta}}_{0s}(\lambda)$) is null, in the literature it is common to handle this parameter separately from all other coefficients ($\boldsymbol{\beta}_{0s}$) (Christensen 2011).

Note that

$$
E\left[\widehat{\boldsymbol{\beta}}_{0s}(\lambda)\right] = \boldsymbol{\Gamma}_s(\boldsymbol{\Lambda}_s + \lambda\boldsymbol{I}_p)^{-1}\boldsymbol{\Lambda}_s\boldsymbol{\beta}_{0s}^*
$$

and

$$\mathrm{Var}\left(\widehat{\beta}_{0s}(\lambda)\right) = \boldsymbol{\Gamma}_s(\boldsymbol{\Lambda}_p + \lambda \boldsymbol{I}_p)^{-1}\boldsymbol{\Lambda}_s(\boldsymbol{\Lambda}_p + \lambda \boldsymbol{I}_p)^{-1}\boldsymbol{\Gamma}_s^{\mathrm{T}},$$

where $\boldsymbol{X}_{1s}^{\mathrm{T}}\boldsymbol{X}_{1s} = \boldsymbol{\Gamma}_s\boldsymbol{\Lambda}_s\boldsymbol{\Gamma}_s^{\mathrm{T}}$ is the spectral decomposition of $\boldsymbol{X}_{1s}^{\mathrm{T}}\boldsymbol{X}_{1s}$ and $\boldsymbol{\beta}_{0s}^{*} = \boldsymbol{\Gamma}_s^{\mathrm{T}}\boldsymbol{\beta}_{0s}$. So the conditional expected prediction error at $\boldsymbol{x}_o$ when using the Ridge solution is

$$\begin{aligned}
\mathrm{EPE}_\lambda(\boldsymbol{x}_o) &= E_{Y,Y_o|\boldsymbol{X},\boldsymbol{x}_o}\left[\left(Y_o - \boldsymbol{x}_o^{*\mathrm{T}}\widehat{\boldsymbol{\beta}}_s^R(\lambda)\right)^2\right] \\
&= E_{Y,Y_o|\boldsymbol{X},\boldsymbol{x}_o}\left[\left(Y_o - E(Y_o|\boldsymbol{x}_0) + E(Y_o|\boldsymbol{x}_0) - \boldsymbol{x}_o^{*\mathrm{T}}\widehat{\boldsymbol{\beta}}_s^R(\lambda)\right)^2\right] \\
&= \sigma^2 + E_{Y,Y_o|\boldsymbol{X},\boldsymbol{x}_o}\left[\left(\boldsymbol{x}_o^{*\mathrm{T}}\boldsymbol{\beta}_s - \boldsymbol{x}_o^{*\mathrm{T}}\widehat{\boldsymbol{\beta}}_s^R(\lambda)\right)^2\right] \\
&= \sigma^2 + \left[\left(\boldsymbol{x}_o^{*\mathrm{T}}\boldsymbol{\beta}_s - \boldsymbol{x}_o^{*\mathrm{T}}E_{Y|\boldsymbol{X}}\left(\widehat{\boldsymbol{\beta}}_s^R(\lambda)\right)\right)^2\right] + \mathrm{Var}\left(\boldsymbol{x}_o^{*\mathrm{T}}\widehat{\boldsymbol{\beta}}_s^R(\lambda)|\boldsymbol{x}_o\right) \\
&= \sigma^2 + \left[\left(\mu + \boldsymbol{x}_o^{\mathrm{T}}\boldsymbol{\beta}_{0s} - \mu - \boldsymbol{x}_o^{\mathrm{T}}\boldsymbol{\Gamma}_s(\boldsymbol{\Lambda}_s + \lambda\boldsymbol{I}_p)^{-1}\boldsymbol{\Lambda}_s\boldsymbol{\beta}_{0s}^{*}\right)^2\right] + \sigma^2\left[\frac{1}{n} + \boldsymbol{x}_o^{\mathrm{T}}\boldsymbol{\Gamma}_s(\boldsymbol{\Lambda}_p + \lambda\boldsymbol{I}_p)^{-1}\boldsymbol{\Lambda}_s(\boldsymbol{\Lambda}_p + \lambda\boldsymbol{I}_p)^{-1}\boldsymbol{\Gamma}_s^{T}\boldsymbol{x}_o\right] \\
&= \sigma^2 + \left[\left(\boldsymbol{x}_o^{**\mathrm{T}}\boldsymbol{\beta}_{0s}^{*} - \boldsymbol{x}_o^{**\mathrm{T}}(\boldsymbol{\Lambda}_s + \lambda\boldsymbol{I}_p)^{-1}\boldsymbol{\Lambda}_s\boldsymbol{\beta}_{0s}^{*}\right)^2\right] + \sigma^2\left[\frac{1}{n} + \boldsymbol{x}_o^{**\mathrm{T}}(\boldsymbol{\Lambda}_p + \lambda\boldsymbol{I}_p)^{-1}\boldsymbol{\Lambda}_s(\boldsymbol{\Lambda}_p + \lambda\boldsymbol{I}_p)^{-1}\boldsymbol{x}_o^{**}\right] \\
&= \sigma^2 + \left[\sum_{j=1}^{p}\left(1 - \frac{\lambda_j}{\lambda_j + \lambda}\right)x_{oj}^{**}\beta_{js}^{*}\right]^2 + \sigma^2\left(\frac{1}{n} + \sum_{j=1}^{p}\frac{\lambda_j}{(\lambda_j + \lambda)^2}\left(x_{oj}^{**}\right)^2\right),
\end{aligned}$$

where $\boldsymbol{x}_o^{**} = \boldsymbol{\Gamma}_s^{\mathrm{T}}\boldsymbol{x}_o^{*} = \left(x_{o0}^{**}, \ldots, x_{op}^{**}\right)^{\mathrm{T}}$ and $\boldsymbol{\beta}_{0s}^{*} = \boldsymbol{\Gamma}_s^{\mathrm{T}}\boldsymbol{\beta}_{0s} = \left(\beta_{1s}^{*}, \ldots, \beta_{ps}^{*}\right)^{\mathrm{T}}$. The second and third terms of the last equality correspond to the squared bias and the variance of $\boldsymbol{x}_o^{*\mathrm{T}}\widehat{\boldsymbol{\beta}}_s^R(\lambda)$ as an estimator of $\boldsymbol{x}_o^{*\mathrm{T}}\boldsymbol{\beta}_s$, respectively. By setting $\lambda = 0$, this EPE corresponds to the EPE of the OLS prediction but with standardized variables, while by letting λ be very large, the variance will decrease and the squared bias will increase.

More importantly, because the derivative of $\mathrm{EPE}_\lambda(\boldsymbol{x}_o)$ with respect to λ, $\frac{d}{d\lambda}\mathrm{PE}_\lambda(\boldsymbol{x}_o)$, is a right continuous function at $\lambda = 0$, and for $\boldsymbol{X}_{1s}$ of full column rank, $\lim_{\lambda \to 0^+}\frac{d}{d\lambda}\mathrm{PE}_\lambda(\boldsymbol{x}_o) = -2\sigma^2\sum_{j=1}^{p}\frac{\left(x_{oj}^{**}\right)^2}{\lambda_j^2} = c$; then for $\epsilon = -\frac{c}{2} > 0$, we have that $\lambda^* > 0$ such that $\left|\frac{d}{d\lambda}\mathrm{PE}_\lambda(\boldsymbol{x}_o) - c\right| < \epsilon$ for $\lambda < \lambda^*$. From this we have that $\frac{d}{d\lambda}\mathrm{PE}_\lambda(\boldsymbol{x}_o) < -\frac{c}{2} + c = \frac{c}{2} < 0$ for all $\lambda < \lambda^*$, for some $\lambda^* > 0$. Then, at least in the interval $[0, \lambda^*]$, the expected prediction error at $\boldsymbol{x}_o$ shows a decreasing behavior, which indicates that there is a value of λ such that with the Ridge regression estimation of beta coefficients, we can get a smaller prediction error than with the OLS prediction. Figure 3.3 shows a graphic representation of this behavior of Ridge prediction, where the lower EPE is reached at about $\lambda = \exp(2.22)$. Figure 3.3 also shows the increasing and decreasing behavior of the bias-squared and the variance involved.

When $\boldsymbol{X}_{1s}$ is not full column rank, the previous argument regarding the behavior of the EPE of the Ridge solution is already not valid directly, but it could be used for

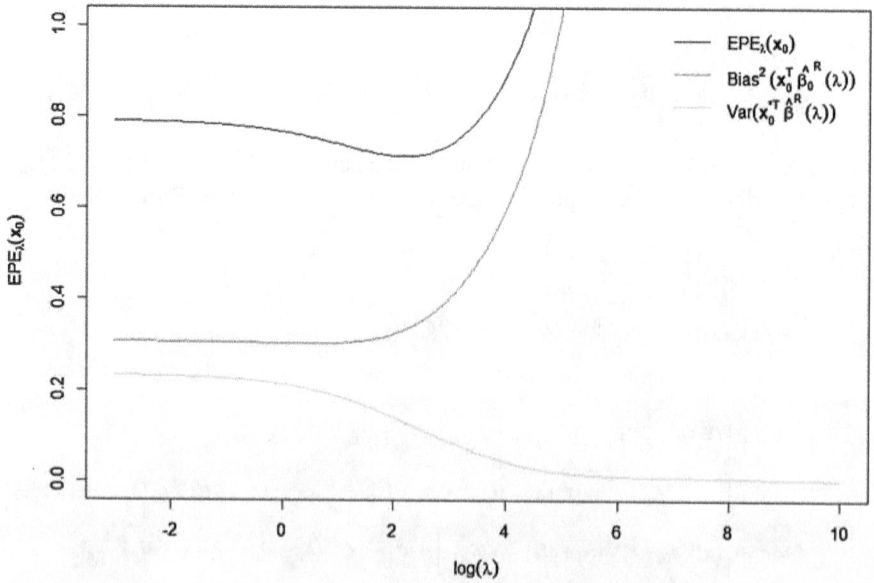

Fig. 3.3 Behavior of the expected prediction error at $\boldsymbol{x}_o$ of the Ridge solution

validating part of the more general case. To see this, first note that the spectral decomposition of $\boldsymbol{X}_{1s}^{\mathrm{T}}\boldsymbol{X}_{1s}$ can be reduced to

$$\boldsymbol{X}_{1s}^{\mathrm{T}}\boldsymbol{X}_{1s} = [\boldsymbol{\Gamma}_{1s}\ \boldsymbol{\Gamma}_{2s}]\begin{bmatrix} \boldsymbol{\Lambda}_{1s} & \mathbf{0} \\ \mathbf{0}^{\mathrm{T}} & \boldsymbol{\Lambda}_{2s} \end{bmatrix}\begin{bmatrix} \boldsymbol{\Gamma}_{1s}^{\mathrm{T}} \\ \boldsymbol{\Gamma}_{2s}^{\mathrm{T}} \end{bmatrix} = \boldsymbol{\Gamma}_{1s}^{\mathrm{T}}\boldsymbol{\Lambda}_{1s}\boldsymbol{\Gamma}_{1s},$$

where $\boldsymbol{\Lambda}_{1s} = \mathrm{Diag}(\lambda_1, \ldots, \lambda_{p^*}), p^* =$ rank $(\mathbf{X}_{1c})$ is the rank of design matrix and $\boldsymbol{\Lambda}_{2s}$ is the null matrix of order $(p - p^*) \times (p - p^*)$. Furthermore, because $\boldsymbol{\Gamma}_s^{\mathrm{T}}\boldsymbol{\Gamma}_s = \boldsymbol{I}_p$ implies that $\boldsymbol{\Gamma}_{2s}^{\mathrm{T}}\boldsymbol{X}_{1s}^{\mathrm{T}}\boldsymbol{X}_{1s}\boldsymbol{\Gamma}_{2s} = \mathbf{0}$, which in turn implies that $\boldsymbol{X}_{1s}\boldsymbol{\Gamma}_{2s} = \mathbf{0}$, then the MLR can be conveniently expressed by

$$\begin{aligned} \boldsymbol{y} &= \mathbf{1}_n\mu + \boldsymbol{X}_{1s}\boldsymbol{\beta}_{0s} + \epsilon \\ &= \mathbf{1}_n\mu + \boldsymbol{X}_{1s}\boldsymbol{\Gamma}_s\boldsymbol{\Gamma}_s^{\mathrm{T}}\boldsymbol{\beta}_{0s} + \epsilon \\ &= \mathbf{1}_n\mu + \boldsymbol{X}_{1s}[\boldsymbol{\Gamma}_{1s}\ \boldsymbol{\Gamma}_{2s}]\boldsymbol{\Gamma}_s^{\mathrm{T}}\boldsymbol{\beta}_{0s} + \epsilon \\ &= \mathbf{1}_n\mu + [\boldsymbol{X}_{1s}\boldsymbol{\Gamma}_{1s}\ \boldsymbol{X}_{1s}\boldsymbol{\Gamma}_{2s}]\boldsymbol{\beta}_{0s}^* + \epsilon \\ &= \mathbf{1}_n\mu + \boldsymbol{X}_{1s}^*\boldsymbol{\beta}_{01s}^* + \epsilon, \end{aligned}$$

where $\boldsymbol{\beta}_{0s}^* = \boldsymbol{\Gamma}_s^{\mathrm{T}}\boldsymbol{\beta}_{0s}$, $\boldsymbol{X}_{1s}^* = \boldsymbol{X}_{1s}\boldsymbol{\Gamma}_{1s}$, and $\boldsymbol{\beta}_{0s}^* = \boldsymbol{\Gamma}_s^{\mathrm{T}}\boldsymbol{\beta}_{0s} = [\boldsymbol{\beta}_{0s}^{T}\boldsymbol{\Gamma}_{1s}, \boldsymbol{\beta}_{0s}^{\mathrm{T}}\,\boldsymbol{\Gamma}_{2s}\,]^{\mathrm{T}} = [\boldsymbol{\beta}_{01s}^{*}{}^{\mathrm{T}}\ \boldsymbol{\beta}_{02s}^{*\mathrm{T}}]^{\mathrm{T}}$.

Also, from similar arguments, note that the penalized residual sum of squares of the Ridge solution can be expressed by

$$\begin{aligned}
&(\boldsymbol{y}-1_n\mu-\boldsymbol{X}_{1s}\boldsymbol{\beta}_{0s})^{\mathrm{T}}(\boldsymbol{y}-1_n\mu-\boldsymbol{X}_{1s}\boldsymbol{\beta}_{0s})+\lambda\boldsymbol{\beta}_{0s}^{\mathrm{T}}\boldsymbol{\beta}_{0s}\\
&=\left(\boldsymbol{y}-1_n\mu-\boldsymbol{X}_{1s}^{*}\boldsymbol{\beta}_{01s}^{*}\right)^{\mathrm{T}}\left(\boldsymbol{y}-1_n\mu-\boldsymbol{X}_{1s}^{*}\boldsymbol{\beta}_{01s}^{*}\right)+\lambda\boldsymbol{\beta}_{0s}^{\mathrm{T}}\left(\boldsymbol{\Gamma}_{1s}^{\mathrm{T}}\boldsymbol{\Gamma}_{1s}+\boldsymbol{\Gamma}_{2s}^{\mathrm{T}}\boldsymbol{\Gamma}_{2s}\right)\boldsymbol{\beta}_{0s}\\
&=\left(\boldsymbol{y}-1_n\mu-\boldsymbol{X}_{1s}^{*}\boldsymbol{\beta}_{01s}^{*}\right)^{\mathrm{T}}\left(\boldsymbol{y}-1_n\mu-\boldsymbol{X}_{1s}^{*}\boldsymbol{\beta}_{01s}^{*}\right)+\lambda\boldsymbol{\beta}_{01s}^{*\mathrm{T}}\boldsymbol{\beta}_{01s}^{*}+\lambda\boldsymbol{\beta}_{02s}^{*\mathrm{T}}\boldsymbol{\beta}_{02s}^{*}
\end{aligned}$$

This function of $\boldsymbol{\beta}_{0s}^{*}$ is minimized at $\widetilde{\boldsymbol{\beta}}_{0s}^{*}(\lambda)=\left(\widetilde{\boldsymbol{\beta}}_{01s}^{*\mathrm{T}}(\lambda),\boldsymbol{0}_{p-p^*}^{\mathrm{T}}\right)^{\mathrm{T}}$, where $\widetilde{\boldsymbol{\beta}}_{01s}^{*\mathrm{T}}(\lambda)=\left(\boldsymbol{\Lambda}_{1s}+\lambda\boldsymbol{I}_{p^*}\right)^{-1}\boldsymbol{\Gamma}_{1s}^{\mathrm{T}}\boldsymbol{X}_{1s}^{\mathrm{T}}\boldsymbol{y}$ is the Ridge solution of the MLR expressed in terms of $\boldsymbol{X}_{1s}^{*}\boldsymbol{\beta}_{0s}^{*}$. Furthermore, because $\boldsymbol{\beta}_{0s}^{*}=\boldsymbol{\Gamma}_{s}^{\mathrm{T}}\boldsymbol{\beta}_{0s}$ is a non-singular transformation, the original Ridge solution of $\boldsymbol{\beta}_s$ can be expressed in terms of $\widetilde{\boldsymbol{\beta}}_{0s}^{*}(\lambda)$ as $\widehat{\boldsymbol{\beta}}_s(\lambda)=\left(\bar{y}_n,\widehat{\boldsymbol{\beta}}_{0s}^{*\mathrm{T}}(\lambda)\right)^{\mathrm{T}}$, where $\widehat{\boldsymbol{\beta}}_{0s}(\lambda)=\boldsymbol{\Gamma}_s\widehat{\boldsymbol{\beta}}_{0s}^{*\mathrm{T}}(\lambda)=\boldsymbol{\Gamma}_{1s}\widetilde{\boldsymbol{\beta}}_{01s}^{*}(\lambda)$. Then, in a similar fashion as before, the conditional expected prediction error at $\boldsymbol{x}_o$ by using the Ridge solution in this case can be computed as

$$\begin{aligned}
&\mathrm{EPE}_{\lambda}(\boldsymbol{x}_o)\\
&=\sigma^2+E_{Y,Y_o|X,\boldsymbol{x}_o}\left[\left(\boldsymbol{x}_o^{*\mathrm{T}}\boldsymbol{\beta}_s-\boldsymbol{x}_o^{*\mathrm{T}}\widehat{\boldsymbol{\beta}}_s^{R}(\lambda)\right)^2\right]\\
&=\sigma^2+\left[\left(\boldsymbol{x}_o^{*\mathrm{T}}\boldsymbol{\beta}_s-\boldsymbol{x}_o^{*\mathrm{T}}E_{Y|X}\left(\widehat{\boldsymbol{\beta}}_s^{R}(\lambda)\right)\right)^2\right]+\mathrm{Var}\left(\boldsymbol{x}_o^{*\mathrm{T}}\widehat{\boldsymbol{\beta}}_s^{R}(\lambda)|\boldsymbol{x}_o\right)\\
&=\sigma^2+\left[\left(\mu+\boldsymbol{x}_o^{\mathrm{T}}\boldsymbol{\Gamma}_s\boldsymbol{\Gamma}_s^{\mathrm{T}}\boldsymbol{\beta}_{0s}-\mu-\boldsymbol{x}_o^{\mathrm{T}}\boldsymbol{\Gamma}_{1s}E_{Y|X}\left(\widehat{\boldsymbol{\beta}}_{0s}^{*\mathrm{T}}(\lambda)\right)\right)^2\right]+\left[\frac{\sigma^2}{n}+\mathrm{Var}\left(\boldsymbol{x}_o^{\mathrm{T}}\boldsymbol{\Gamma}_s\widehat{\boldsymbol{\beta}}_{0s}^{*\mathrm{T}}(\lambda)|\boldsymbol{x}_o\right)\right]\\
&=\sigma^2+\left[\left(\mu+\boldsymbol{x}_o^{\mathrm{T}}\boldsymbol{\Gamma}_s\boldsymbol{\Gamma}_s^{\mathrm{T}}\boldsymbol{\beta}_{0s}-\mu-\boldsymbol{x}_o^{\mathrm{T}}\boldsymbol{\Gamma}_{1s}(\boldsymbol{\Lambda}_{1s}+\lambda\boldsymbol{I}_{p^*})^{-1}\boldsymbol{\Gamma}_{1s}^{\mathrm{T}}\boldsymbol{X}_{1c}^{\mathrm{T}}\boldsymbol{X}_{1c}\boldsymbol{\beta}_{0S}\right)^2\right]\\
&\quad+\sigma^2\left[\frac{1}{n}+\boldsymbol{x}_o^{\mathrm{T}}\boldsymbol{\Gamma}_{1s}(\boldsymbol{\Lambda}_{1s}+\lambda\boldsymbol{I}_{p^*})^{-1}\boldsymbol{\Lambda}_{1s}(\boldsymbol{\Lambda}_{1s}+\lambda\boldsymbol{I}_{p^*})^{-1}\boldsymbol{\Gamma}_{1s}^{T}\boldsymbol{x}_o\right]\\
&=\begin{cases}\sigma^2+\left[\left(\sum_{j=1}^{p^*}\left(1-\frac{\lambda_j}{\lambda_j+\lambda}\right)x_{oj}^{*}\beta_{oj}^{*}\right)^2\right]+\sigma^2\left[\frac{1}{n}+\sum_{j=1}^{p^*}\frac{\lambda_j}{(\lambda_j+\lambda)^2}\left(x_{oj}^{*}\right)^2\right] & \text{if } x_o=\boldsymbol{\Gamma}_{1s}\boldsymbol{a}_1\\
\sigma^2+\left[\boldsymbol{x}_o^{T}\boldsymbol{\beta}_{0s}\right]^2+\frac{\sigma^2}{n} & \text{if } x_o=\boldsymbol{\Gamma}_{2s}\boldsymbol{a}_2,\end{cases}
\end{aligned}$$

where $\boldsymbol{x}_o^{*}=\boldsymbol{\Gamma}_s^{\mathrm{T}}\boldsymbol{x}_o=\left[x_{o1}^{*},\ldots,x_{op}^{*}\right]^{\mathrm{T}}$, $\boldsymbol{\beta}_o^{*}=\boldsymbol{\Gamma}_s^{\mathrm{T}}\boldsymbol{\beta}_{os}=\left[\beta_{o1}^{*},\ldots,\beta_{op}^{*}\right]^{\mathrm{T}}$, and $\boldsymbol{a}_1\in\mathbb{R}^{p^*}$ and $\boldsymbol{a}_2\in\mathbb{R}^{p-p^*}$. So, using a similar argument as before, in the first case ($x_o=\boldsymbol{\Gamma}_{1s}\boldsymbol{a}_1$), the value of $\lambda>0$ is such that the expected prediction error at $\boldsymbol{x}_o$ is better than that obtained with the OLS approach, $\widehat{\boldsymbol{\beta}}_s(0)=\lim_{\lambda\to0}\widehat{\boldsymbol{\beta}}_s(\lambda)=\left[\bar{y}_n,\left(\boldsymbol{\Gamma}_{1s}\Lambda_{1s}^{-1}\boldsymbol{\Gamma}_{1s}^{\mathrm{T}}\boldsymbol{X}_{1c}^{\mathrm{T}}\boldsymbol{y}\right)^{\mathrm{T}}\right]^{\mathrm{T}}$. In the second case, $x_o=\boldsymbol{\Gamma}_{2s}\boldsymbol{a}_2$, the EPE($\boldsymbol{x}_o$) in both approaches is the same and doesn't depend on λ, so in such cases, no improved gain with regard to the Ridge solution is achieved. A third case was included, that is, when the target feature is of the form $\boldsymbol{x}_o=\boldsymbol{\Gamma}_{1s}\boldsymbol{a}_1+\boldsymbol{\Gamma}_{2s}\boldsymbol{a}_2$. In this case, under the described argument, the advantage of Ridge regression over the OLS approach in a prediction context is not clear.

However, in practice, we don't know the true value of the parameters, and we need to evaluate the test error in all possible values of the training sample, which we also don't have. So a common way to choose the λ value is by cross-validation. For more details about validation strategies, see Chap. 4. For example, with a k-fold CV, the complete data set is divided into K balanced disjoint subsets, S_k, $k = 1, \ldots, K$. One subset is used as validation and the rest are used to fit the model in each value of a chosen grid of values of λ. This procedure is repeated K times, where each time a subset in the partition is taken as the validation set. A more detailed k-fold CV procedure is described below:

1. First, choose a grid of values of λ, $\lambda = (\lambda_1, \ldots, \lambda_L)$.
2. Remove the subset S_k and for each value λ_l in the grid, fit the model with the remaining $K - 1$ elements of the partition denoted by $\widehat{\boldsymbol{\beta}}^R_{-k}(\lambda_l)$, the corresponding Ridge estimation of $\boldsymbol{\beta}$, and compute the average prediction error across all observations in the validation set S_k as

$$\widehat{\text{APE}}_{-k}(\lambda_l) = \frac{1}{|S_k|} \sum_{y_i \in S_k} \left(y_i - \boldsymbol{x}_i^{*\text{T}} \widehat{\boldsymbol{\beta}}^R_{-k}(\lambda_l) \right)^2,$$

   where $|S_k|$ denotes the total observations in partition k.
3. Choose as the best value of λ in the grid $\left(\widetilde{\lambda}^*\right)$, the one with the lower average prediction error across all partitions, that is

$$\widetilde{\lambda}^* = \arg\min_{\lambda_l} \widehat{\text{APE}}(\lambda_l),$$

   where $\widehat{\text{APE}}(\lambda_l) = \frac{1}{K} \sum_{k=1}^{K} \widehat{\text{APE}}_{-k}(\lambda_l)$.
4. Once $\widetilde{\lambda}^*$ is chosen, we fit the model with the complete data set and the prediction of new individuals with feature $\boldsymbol{x}_o$ can be made with $\widehat{y}_i = \boldsymbol{x}_o^{*T} \widehat{\boldsymbol{\beta}}^R\left(\widetilde{\lambda}^*\right)$, where $\widehat{\boldsymbol{\beta}}^R\left(\widetilde{\lambda}^*\right)$ is the Ridge estimation of $\boldsymbol{\beta}$ at $\lambda = \widetilde{\lambda}^*$.

It is important to point out that very often the performance of the model needs to be evaluated for comparison purposes with other competing models. A common way to do this is to split the data set several times into two subsets, one for training the model (D_{tr}) (to fit the model) and the other for testing (D_{tst}) it, in which the predictive ability of a model is tested. In each splitting, only the training data set (D_{tr}) is used to train the model (by steps 1–3 before fitting the whole training data set), and the prediction evaluation of the fitted model is made with the testing data set, as explained before in point 4. The prediction evaluation of the testing data set is done by an empirical "estimate" of the EPE, $\text{MSE} = \frac{1}{|D_{\text{tst}}|} \sum_{i \in D_{\text{tst}}} \left(y_i - \boldsymbol{x}_o^{*\text{T}} \widehat{\boldsymbol{\beta}}^R\left(\widetilde{\lambda}^*\right), \right)^2$, and finally, an average evaluation of the performance of the model is obtained across all chosen splittings. See Chap. 4 for more explicit details.

The Ridge solution can also be obtained from a Bayesian formulation. To do this, consider the MLR model described before with standardized features and the vector of residuals distributed as $N_n(\mathbf{0}, \sigma^2 \boldsymbol{I}_n)$. With this assumption, the vector of responses $\boldsymbol{y}$ is distributed in a multivariate normal distribution with vector mean $\mathbf{1}_n\mu + \boldsymbol{X}_{1s}\boldsymbol{\beta}_{0s}$ and variance–covariance matrix $\sigma^2 \boldsymbol{I}_n$. Then, to complete the Bayesian formulation, assume $\boldsymbol{\beta}_{os} \sim N_p\left(\mathbf{0}, \sigma_\beta^2 \boldsymbol{I}_p\right)$ as the prior distribution of the beta coefficients in $\boldsymbol{\beta}_{0s}$ and a "flat" prior for the intercept μ, where σ^2 and σ_β^2 are known. Under this Bayesian specification, the posterior distribution of $\boldsymbol{\beta}_s$ is

$$
\begin{aligned}
& f(\boldsymbol{\beta}_s | \boldsymbol{y}, \boldsymbol{X}_{1s}) \\
& \propto \exp\left[-\frac{1}{2\sigma^2}(\boldsymbol{y} - \boldsymbol{X}_s\boldsymbol{\beta}_s)^{\mathrm{T}}(\boldsymbol{y} - \boldsymbol{X}_s\boldsymbol{\beta}_s)\right] \exp\left(-\frac{1}{2\sigma_\beta^2}\boldsymbol{\beta}_{0s}^{\mathrm{T}}\boldsymbol{\beta}_{0s}\right) \\
& \propto \exp\left\{-\frac{1}{2}\left[\boldsymbol{\beta}_s^{\mathrm{T}}\left(\sigma_\beta^{-2}\boldsymbol{D} + \sigma^{-2}\boldsymbol{X}_s^{\mathrm{T}}\boldsymbol{X}_s\right)\boldsymbol{\beta}_s - 2\sigma^{-2}\boldsymbol{y}^{\mathrm{T}}\boldsymbol{X}_s\boldsymbol{\beta}_s\right]\right\} \\
& \propto \exp\left\{-\frac{1}{2}\left(\boldsymbol{\beta}_s - \widetilde{\boldsymbol{\beta}}_s\right)^{\mathrm{T}}\widetilde{\boldsymbol{\Sigma}}_\beta^{-1}\left(\boldsymbol{\beta}_s - \widetilde{\boldsymbol{\beta}}_s\right)\right\},
\end{aligned}
$$

where $\widetilde{\boldsymbol{\Sigma}}_\beta = \left(\sigma_\beta^{-2}\boldsymbol{D} + \sigma^{-2}\boldsymbol{X}_s^{\mathrm{T}}\boldsymbol{X}_s\right)^{-1} = \sigma^2\left(\sigma^2/\sigma_\beta^2\boldsymbol{D} + \boldsymbol{X}_s^{\mathrm{T}}\boldsymbol{X}_s\right)^{-1}$, $\widetilde{\boldsymbol{\beta}}_s = \sigma^{-2}\widetilde{\boldsymbol{\Sigma}}_\beta\boldsymbol{X}_s^{\mathrm{T}}\boldsymbol{y}$, and $\boldsymbol{D}$ is the diagonal penalty matrix. That is, the posterior distribution of $\boldsymbol{\beta}_s$ is a multivariate normal distribution with vector mean $\widetilde{\boldsymbol{\beta}}_s = \sigma^{-2}\widetilde{\boldsymbol{\Sigma}}_\beta\boldsymbol{X}_s^{\mathrm{T}}\boldsymbol{y} = \left(\sigma^2/\sigma_\beta^2\boldsymbol{D} + \boldsymbol{X}_s^{\mathrm{T}}\boldsymbol{X}_s\right)^{-1}\boldsymbol{X}_s^{\mathrm{T}}\boldsymbol{y}$ and variance–covariance matrix $\widetilde{\boldsymbol{\Sigma}}_\beta = \sigma^2\left(\sigma^2/\sigma_\beta^2\boldsymbol{D} + \boldsymbol{X}_s^{\mathrm{T}}\boldsymbol{X}_s\right)^{-1}$. Then, by taking $\lambda = \sigma^2/\sigma_\beta^2$, we have that the mean/mode of the posterior distribution of $\boldsymbol{\beta}_s$ coincides with the Ridge estimation described before, $\widetilde{\boldsymbol{\beta}}^R(\lambda)$.

Example 2 We considered a genomic example to illustrate the Ridge regression approach and the CV process to choose the learning parameter λ (WheatMadaToy, PH the response). This data set consists of 50 observations corresponding to 50 lines and a relationship genomic matrix computed from marker information. Table 3.1 shows the prediction behavior of the Ridge and the OLS approaches in terms of the MSEP, across five different splittings obtained by partitioning the complete data set into five subsets: the data of a subset are used as a testing set and the rest to train the model. For training the model, a five-fold cross-validation (5FCV) was used along the lines following steps 1–3 described before, and the prediction performance was done following step 4.

Table 3.1 Prediction behavior of the Ridge and OLS regression models across different partitions of the complete data set: one subset of the partition (20%) is used for evaluating the performance of the model and the rest (80%) for training the model. RR denotes Ridge regression method

Partition	MSE RR	MSE OLS
1	325.40	379.33
2	433.19	454.37
3	803.76	1341.35
4	319.53	312.81
5	403.62	555.10
Average	457.10	608.59

Table 3.1 indicates that in four out of five partitions, the Ridge regression shows less MSE than the corresponding OLS approach. In all these cases, the MSE of the OLS was, on average, 31.46% greater than that of the Ridge regression approach, and in general, on average, by 31.14% (MSE = 421.8834 for Ridge and MSE = 655.8596 for OLS). From this, we have that the Ridge regression approach shows a better prediction performance than the OLS. The large variation of the MSE between folds observed in this example could indicate that for obtaining a more precise comparison between models, a larger number of partitions need to be used. Often, the use of more partitions is avoided when larger data sets are used in applications.

The R code used for obtaining this result is the following:

```
#######################R code for Example 2 ###########################
rm(list=ls())
library(BMTME)
data("WheatMadaToy")
dat_F = phenoMada
dim(dat_F)
dat_F$GID = as.character(dat_F$GID)
G = genoMada
eig_G = eigen(G)
G_0.5 = eig_G$vectors%*%diag(sqrt(eig_G$values))%*%t(eig_G$vectors)
X = G_0.5
y = dat_F$PH
n = length(y)
source('TR_RR.R')
#5FCV
set.seed(3)
K = 5
Tab = data.frame()
Grpv = findInterval(cut(sample(1:n,n),breaks=K),1:n)
for(i in 1:K)
{
 Pos_tr = which(Grpv!=i)
 y_tr = y[Pos_tr]
 X_tr = X[Pos_tr,]
 TR_RR = Tr_RR_f(y_tr,X_tr,K=5,KG=100,KR=1)
 lambv = TR_RR$lambv
```

```
  #Tst
  y_tst = y[-Pos_tr] ; X_tst = X[-Pos_tr,]
  #RR
  Pred_RR = Pred_RR_f(y_tst,X_tst,TR_RR)
  #OLS
  Pred_ols = Pred_ols_f(y_tst,X_tst,y_tr,X_tr)
  Tab = rbind(Tab,data.frame(Sim=i,MSEP_RR = Pred_RR$MSEP,
                  MSEP_ols = Pred_ols$MSEP))
  cat('i = ', i,'\n')
}
Tab
```

Tr_RR_f, Pred_RR_f, and Pred_ols_f are R functions accessed by the command source('TR_RR.R'), where TR_RR.R is the file R script defined in Appendix 1.

From the last code, three things are important to point out:

1. The Grpv contains the information of the K=5 folds for the outer CV implemented and each time the model is trained with $K-1$ and tested with the remaining fold.
2. The function that trains the model under Ridge regression is called Tr_RR_f, while the function that obtains the predictions of the testing set of this trained model is Pred_RR_f; both functions are fully described in Appendix 1.
3. The predictions under the OLS method are obtained with the function Pred_ols_f, which is also fully detailed in Appendix 1. It is important to point out that the function Tr_RR_f internally implements an inner k-fold CV to tune the hyperparameter λ required in Ridge regression.

Example 3 (Simulation) To get a better idea about the behavior of the Ridge solution, here we report the results of a small simulation study in a scenario where the number of observations (n=100) is less than the number of features ($p=500$) and these are moderately correlated. Specifically, we generated 100 data sets, each of size 100, from the following model:

$$y_i = 5 + \boldsymbol{x}_i^{\mathrm{T}}\boldsymbol{\beta}_0 + \epsilon_i,$$

where the vector of beta coefficients ($\boldsymbol{\beta}_0$) was set to the values shown in Fig. 3.4, and the features of all the individuals in each data set were generated from a multivariate normal distribution centered on the null vector and variance–covariance matrix $\boldsymbol{\Sigma} = 0.25\boldsymbol{I}_p + 0.75\boldsymbol{J}_p$, where $\boldsymbol{I_p}$ and $\boldsymbol{J_p}$ are the identity matrix and matrix of ones of dimension $p \times p$. The random errors (ϵ_i) were simulated from a normal distribution with mean 0 and variance 0.025.

The behavior of the Ridge and OLS solutions across the 100 simulated data sets is shown in Fig. 3.5. The MSE of Ridge regression is located on the x-axis and the corresponding MSE of the OLS is located on the y-axis. On average, the OLS resulted in an MSE equal to 808.81, which is 30.59% larger than the average MSE

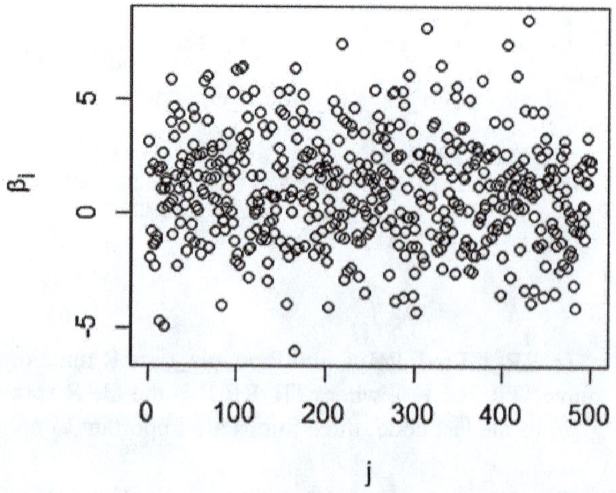

Fig. 3.4 Beta coefficients used in simulation: $\beta_j, j = 1, \ldots, p$

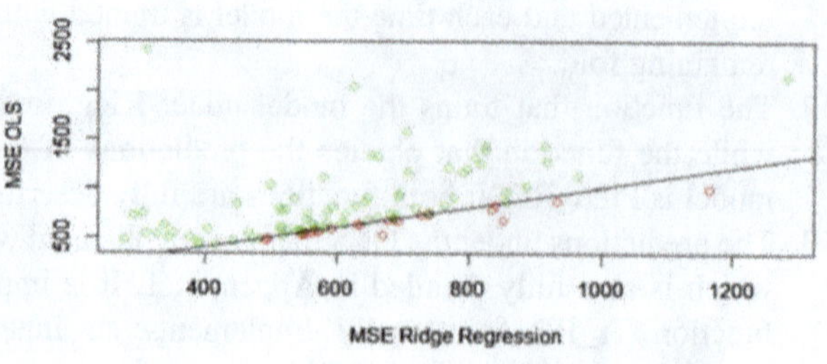

Fig. 3.5 MSE of Ridge regression (MSE RR) versus MSE OLS regression (MSE OLS)

(619.32) of the Ridge approach. In terms of the percentage of simulations in favor of each method, Ridge regression was better in 78 out of 100 simulations, while the OLS was better only in 22 out of 100 simulations. In general, from this small simulation study we obtained more evidence in favor of the Ridge regression method.

The R code used for obtaining this result is the following:

```
###########################R code for Example 3######################
rm(list=ls(all=TRUE))
library(mvtnorm)
library(MASS)
source('TR_RR.R')
set.seed(10)
n = 100
p = 500
Var = 0.25*diag(p)+0.75
Tab = data.frame()
betav = rnorm(p,rpois(p,1),2)
plot(betav,xlab=expression(j),ylab=expression(beta[j]))
for(i in 1:100)
{
 X = rmvnorm(n,rep(0,p),Var)
```

```
dim(X)
y = 5+X%*%betav + rnorm(n,0,0.5)
Pos_tr = sample(1:n,n*0.80)
y_tr = y[Pos_tr] ; X_tr = X[Pos_tr,]
y_tst = y[-Pos_tr] ; X_tst = X[-Pos_tr,]
#Training RR
TR_RR = Tr_RR_f(y_tr,X_tr,K=5,KG=100,KR=1)
TR_RR$lamb_o
lambv = TR_RR$lambv
plot(log(TR_RR$lambv),TR_RR$iPEv_mean)
#Prediction RR in testing data
Pred_RR = Pred_RR_f(y_tst,X_tst,TR_RR)
Pred_RR$MSEP

#OLS
Pred_ols = Pred_ols_f(y_tst,X_tst,y_tr,X_tr)
Pred_ols

Tab = rbind(Tab,data.frame(Sim=i,MSEP_RR = Pred_RR$MSEP,
            MSEP_ols = Pred_ols$MSEP))
cat('i = ', i,'\n')}

Mean_v = colMeans(Tab)
(Mean_v[3]-Mean_v[2])/Mean_v[2]*100

mean(Tab$MSEP_RR<Tab$MSEP_ols)
Pos = which(Tab$MSEP_RR<Tab$MSEP_ols)
mean((Tab$MSEP_ols[Pos]-Tab$MSEP_RR[Pos])/Tab$MSEP_RR[Pos])*100
mean((Tab$MSEP_RR[-Pos]-Tab$MSEP_ols[-Pos])/Tab$MSEP_ols[-Pos])
*100

plot(Tab$MSEP_RR, Tab$MSEP_ols,
  col=ifelse(Tab$MSEP_RR<Tab$MSEP_ols,3,2),
  xlab='MSEP RR', ylab='MSEP OLS')
abline(a=0,b=1)
```

The TR_RR.R script file is the same as the one defined in Example 2 in Appendix 1.

3.6.2 *Lasso Regression*

Like Ridge regression, the Lasso regression solves the OLS problem but penalizes the residual sum squared in a slightly different way. With the standardized variables, the Lasso estimator of $\boldsymbol{\beta}_s$ is defined as

$$\widetilde{\boldsymbol{\beta}}_s^L(\lambda) = \arg\min_{\mu,\boldsymbol{\beta}_{0s}} \text{PRSS}_\lambda(\boldsymbol{\beta}_s),$$

where now $\text{PRSS}_\lambda(\boldsymbol{\beta}_s) = \sum_{i=1}^{n}\left(y_i - \mu - \sum_{j=1}^{p} x_{ijs}\beta_{js}\right)^2 + \lambda\sum_{j=1}^{p}|\beta_{js}|$ is the RSS($\boldsymbol{\beta}$) but penalized by the sum of the absolute regression coefficients. For $\lambda = 0$, the solution is the OLS, while when λ is large, the OLS solutions are shrunken toward 0 (Tibshirani 1996).

Note that for any given values of $\boldsymbol{\beta}_{0s}$, the value of μ that minimizes $\text{PRSS}_\lambda(\boldsymbol{\beta}_s)$ is the sample mean of the responses, $\widetilde{\mu} = \frac{1}{n}\sum_{i=1}^{n} y_i$, the same as the Ridge estimator. However, the rest of the Lasso estimator of $\boldsymbol{\beta}_s$, $\boldsymbol{\beta}_{0s}$, cannot be obtained analytically, so numerical methods are often used.

Although there are efficient algorithms for computing the entire regularization path for the Lasso regression coefficients (Efron et al. 2004; Friedman et al. 2008), here we will describe the coordinate-wise descent given in Friedman et al. (2007). The idea of this method is to successively optimize the $\text{PRSS}_\lambda(\boldsymbol{\beta}_s)$ one parameter at a time (beta coefficient). Holding β_{ks}, $j \neq k$, fixed at their current values $\widetilde{\beta}_{js}(\lambda)$, the value of β_k that minimizes $\text{PRSS}_\lambda(\boldsymbol{\beta}_s)$ is given by

$$\begin{aligned}\widetilde{\beta}^*_{ks}(\lambda) &= S\left(\sum_{i=1}^{n} x_{ijs}\left(y_i - \widetilde{y}_i^{(k)}\right), \lambda\right)\\ &= S\left(n\widetilde{\beta}_{ks}(\lambda) + \sum_{i=1}^{n} x_{ijs}(y_i - \widetilde{y}_i), \lambda\right),\end{aligned}$$

where $\widetilde{y}_i^{(k)} = \bar{y} + \sum_{\substack{j=1\\ j\neq k}}^{p} x_{ijs}\widetilde{\beta}_{js}(\lambda)$ and $S(\beta, \lambda) = \begin{cases}\beta - \lambda & \text{if } \beta > 0 \text{ and } \lambda < |\beta| \\ \beta + \lambda & \text{if } \beta < 0 \text{ and } \lambda < |\beta| \\ 0 & \text{if } \lambda \geq |\beta|\end{cases}$. To obtain the Lasso estimate of $\boldsymbol{\beta}_{0s}$, this process is repeated across all the coefficients until a convergence threshold criterion is reached.

This algorithm can be implemented with the glmnet R package (Friedman et al. 2010) as part of a more general penalty regression (elastic net), which is defined as a combination of the Ridge and Lasso penalties. Due to the structure of the algorithm, this can be used on very large data sets and can benefit from sparsity in the explanatory variables (Friedman et al. 2008).

Equivalently, the Lasso estimator of beta coefficients $\boldsymbol{\beta}_{0s}$ can be defined as

$$\widetilde{\boldsymbol{\beta}}^L_{0s}(\lambda) = \underset{\boldsymbol{\beta}_{0s}}{\text{argmin}} \sum_{i=1}^{n}\left(y_i - \bar{y} - \sum_{j=1}^{p} x_{ijs}\beta_{js}\right)^2$$

$$\text{subject to} \sum_{j=1}^{p}|\beta_{js}| \leq t$$

With this, a graphic representation of the Lasso estimator is like the Ridge (see Fig. 3.6). The nested ellipsoids correspond to contour plots of RSS($\boldsymbol{\beta}$) and the green

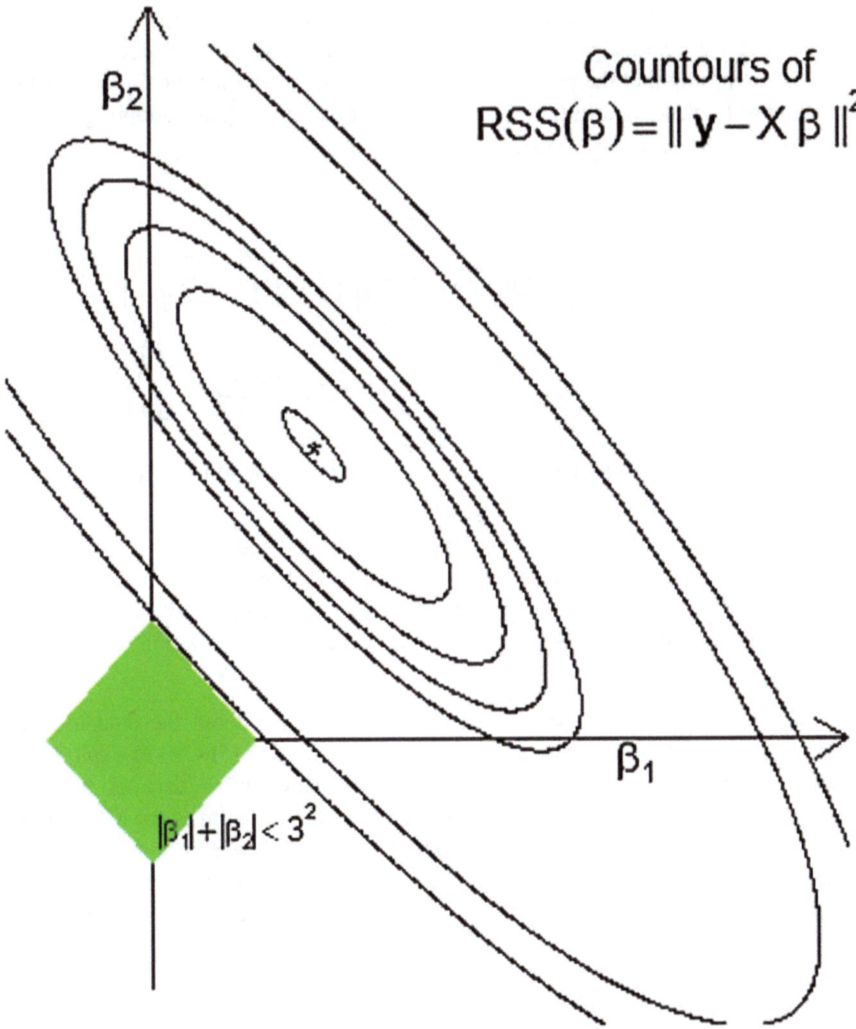

Fig. 3.6 Graphic representation of the Lasso solution of the OLS with restriction $\sum_{j=1}^{p} |\beta_j| < 3^2$. The green region contains the Lasso solution

region is the restriction with $t = 3^2$, which contains the Lasso solution. Indeed, the Lasso solution is the first point of the contours that touches the square, and this will sometimes be in a corner that makes some coefficients zero. Because there are no corners in Ridge regression, this will rarely happen (Tibshirani 1996).

The Lasso estimator can also be derived from a Bayesian perspective. Supposing that the vector of residuals is distributed as $N_n(\mathbf{0}, \sigma^2 \boldsymbol{I}_n)$, like in the Ridge regression case, and assuming that the priors of $\boldsymbol{\beta}_{0s}$ are independent and identically distributed according to Laplace distribution with mean 0 and variance σ_β^2, and adopting a "flat" prior for μ, with known σ^2 and σ_β^2, the posterior distribution of $\boldsymbol{\beta}$ is

$$
\begin{aligned}
& f(\boldsymbol{\beta}_s | \boldsymbol{y}, \boldsymbol{X}_{1s}) \\
& \propto \exp\left[-\frac{1}{2\sigma^2}(\boldsymbol{y}-\boldsymbol{X}_s\boldsymbol{\beta}_s)^{\mathrm{T}}(\boldsymbol{y}-\boldsymbol{X}_s\boldsymbol{\beta}_s)\right]\prod_{i=1}^{n}\exp\left(-\frac{\sqrt{2}}{\sigma_\beta^2}|\beta_{js}|\right) \\
& \propto \exp\left\{-\frac{1}{2\sigma^2}\left[\sum_{i=1}^{n}\left(y_i-\mu-\sum_{j=1}^{p}x_{ijs}\beta_{js}\right)+\lambda\sum_{j=1}^{p}|\beta_{js}|\right]\right\},
\end{aligned}
$$

where $\lambda = \sqrt{8}\sigma^2/\sigma_\beta^2$. Then, the model of the posterior distribution of $\boldsymbol{\beta}_s$ corresponds to the Lasso estimator described before, $\widetilde{\boldsymbol{\beta}}^L(\lambda)$.

The performance of Lasso regression in terms of prediction error is sometimes comparable to Ridge regression (Hastie et al. 2009). However, as we pointed out before, and based on the nature of the restriction term, for any given value of t, only a subset of the coefficients β_{js} is nonzero, so this gives a sparse solution (Efron et al. 2004).

Example 4 To illustrate Lasso regression, here we considered the data used in Example 2, but instead of using a five-fold cross-validation (5FCV) to explore the behavior of this, we built 100 random splittings of the complete data set: 80% for training and 20% for testing. Figure 3.7 presents a representation of the MSE of the Lasso regression (y-axis) and the MSE corresponding to Ridge regression (x-axis). In 81 out of 100 random splittings, the Ridge regression approach gives a better performance, and in this case, on average, the Lasso regression shows an MSE that is 92.13% greater than the Ridge solution. In the other cases, the Ridge was worse, on average, by 30.91%.

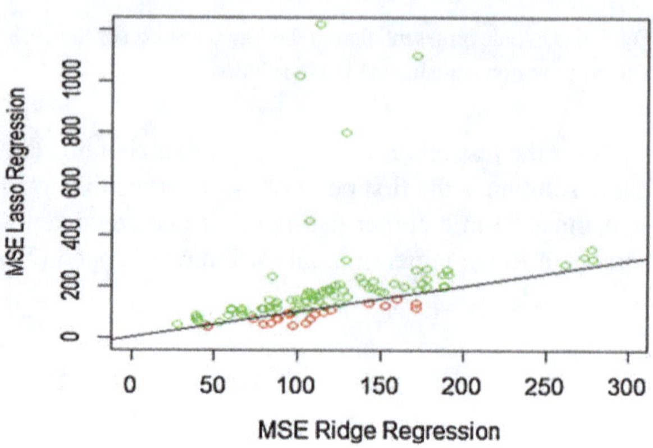

Fig. 3.7 MSE of Ridge regression versus MSE of Lasso regression in 100 random splittings of data: 20% for testing and 80% for training

On average across all the splittings, the performance of the Ridge regression (average = 118.9726 and standard deviation = 50.7193 of MSE) was superior to the Lasso (200.6021 and standard deviation = 222.5494 of MSE) by 68.61%, but this was better than the OLS solution (average = 1609.4635 and standard deviation = 1105.4434 of MSE) by 802.32%, while the Ridge was 1352.80% better than the OLS estimate.

```
########################R code for Example 4########################
rm(list=ls())
library(BMTME)
data("WheatMadaToy")
dat_F = phenoMada
dim(dat_F)
dat_F$GID = as.character(dat_F$GID)
G = genoMada
eig_G = eigen(G)
G_0.5 = eig_G$vectors%*%diag(sqrt(eig_G$values))%*%t(eig_G$vectors)
X = G_0.5
y = dat_F$PH
n = length(y)
library(glmnet)
#5FCV
set.seed(3)
K = 5
Tab = data.frame()
set.seed(1)
for(k in 1:100)
{
 Pos_tr = sample(1:n,n*0.8)
 y_tr = y[Pos_tr]; X_tr = X[Pos_tr,]; n_tr = dim(X_tr)[1]
 y_tst = y[-Pos_tr]; X_tst = X[-Pos_tr,]
 #Partition for internal training the model
 Grpv_k = findInterval(cut(sample(1:n_tr,n_tr),breaks=5),1:n_tr)
 #RR
 A_RR = cv.glmnet(X_tr,y_tr,alpha=0,foldid=Grpv_k,type.
measure='mse')
 yp_RR = predict(A_RR,newx=X_tst,s='lambda.min')
 #LR
 A_LR = cv.glmnet(X_tr,y_tr,alpha=1,foldid=Grpv_k,type.
measure='mse')
 yp_LR = predict(A_LR,newx=X_tst,s='lambda.min')
 #OLS
 A_OLS = glmnet(X_tr,y_tr,alpha=1,lambda=0)
 yp_OLS = predict(A_OLS,newx=X_tst)

 Tab = rbind(Tab,data.frame(PT=k,MSEP_RR = mean((y_tst-yp_RR)^2),
            MSEP_LR = mean((y_tst-yp_LR)^2),
            MSEP_OLS = mean((y_tst-yp_OLS)^2)))
 cat('k = ', k,'\n')
}
Tab
```

Now the key components of the just given R code are

1. Hundred random partitions were implemented where each partition is obtained with Pos_tr = sample(1:n,n*0.8), which means that 80% of the data is used for training and 20% for testing, and for each training set, an inner K=5 fold CV is performed to tune the λ hyperparameter.
2. The Grpv_k contains the information of the K=5 fold inner CV implemented to tune the hyperparameter λ.
3. Now we use the cv.glmnet function that is useful for implementing supervised learning methods with cross-validation. This function belongs to the R package glmnet and the input we give to this function is the training set (X_tr, y_tr), alpha=0, that tells glmnet to implement a Ridge regression method, while alpha=1 orders glmnet to implement a Lasso regression. In foldid=Grpv_k we are given training and testing sets to tune the hyperparameter λ, and in type. measure='mse', we are specifying the metric with which we will evaluate the prediction performance of the inner testing sets to be able to choose the best hyperparameter.
4. The function glmnet with lambda=0 implements the OLS estimator.

It is important to point out that Lasso regression performs particularly well when there is a subset of true coefficients that are small or even zero. It doesn't do as well when all of the true coefficients are moderately large; however, in this case, it can still outperform linear regression over a pretty narrow range of (small) λ values.

3.7 Logistic Regression

The logistic regression is a useful and traditional tool used to explain or predict a binary response based on information of explanatory variables. It models the conditional distribution of the response variable as a Bernoulli distribution with the probability of success given by

$$P(Y_i = 1|\boldsymbol{x}_i) = p(\boldsymbol{x}_i;\boldsymbol{\beta}) = \frac{\exp\left(\beta_0 + \boldsymbol{x}_i^{\mathrm{T}}\boldsymbol{\beta}_0\right)}{1 + \exp\left(\beta_0 + \boldsymbol{x}_i^{\mathrm{T}}\boldsymbol{\beta}_0\right)}.$$

To estimate parameters under logistic regression, suppose that we have a set of data $\left(\boldsymbol{x}_i^{\mathrm{T}}, y_i\right)$, $i = 1, \ldots, n$ (training data), where $\boldsymbol{x}_i = (x_{i1}, \ldots, x_{ip})^{\mathrm{T}}$ is a vector of features measurement and y_i is the response measurement corresponding to the ith drawn individual. To obtain the MLE of $\boldsymbol{\beta}$, first we need to build the likelihood function of the parameters of $\boldsymbol{\beta}$. This is given by

$$\begin{aligned} L(\boldsymbol{\beta};\boldsymbol{y}) &= \prod_i^n p(\boldsymbol{x}_i;\boldsymbol{\beta})^{y_i}[1 - p(\boldsymbol{x}_i;\boldsymbol{\beta})]^{1-y_i} = \prod_i^n \left(\frac{p(\boldsymbol{x}_i;\boldsymbol{\beta})}{1 - p(\boldsymbol{x}_i;\boldsymbol{\beta})}\right)^{y_i} [1 - p(\boldsymbol{x}_i;\boldsymbol{\beta})]^1 \\ &= \exp\left(\sum_{i=1}^n y_i\left(\beta_0 + \boldsymbol{x}_i^{\mathrm{T}}\boldsymbol{\beta}_0\right)\right) \prod_{i=1}^n \frac{1}{1 + \exp\left(\beta_0 + \boldsymbol{x}_i^{\mathrm{T}}\boldsymbol{\beta}_0\right)}. \end{aligned}$$

and from here the log-likelihood is

$$\ell(\boldsymbol{\beta};\boldsymbol{y}) = \log[L(\boldsymbol{\beta};\boldsymbol{y})] = \sum_{i=1}^{n} y_i(\beta_0 + \boldsymbol{x}_i^{\mathrm{T}}\boldsymbol{\beta}_0) - \sum_{i=1}^{n} \log\left[1 + \exp\left(\beta_0 + \boldsymbol{x}_i^{\mathrm{T}}\boldsymbol{\beta}_0\right)\right].$$

Then, because the gradient of the likelihood is given by

$$\begin{aligned}\frac{\partial \ell(\boldsymbol{\beta};\boldsymbol{y})}{\partial \boldsymbol{\beta}} &= \begin{bmatrix} \sum_{i=1}^{n} y_i \\ \sum_{i=1}^{n} y_i x_{i1} \\ \vdots \\ \sum_{i=1}^{n} y_i x_{ip} \end{bmatrix} - \begin{bmatrix} \sum_{i=1}^{n} \frac{\exp(\beta_0 + \boldsymbol{x}_i^{\mathrm{T}}\boldsymbol{\beta}_0)}{1 + \exp(\beta_0 + \boldsymbol{x}_i^{\mathrm{T}}\boldsymbol{\beta}_0)} \\ \sum_{i=1}^{n} \frac{\exp(\beta_0 + \boldsymbol{x}_i^{\mathrm{T}}\boldsymbol{\beta}_0)}{1 + \exp(\beta_0 + \boldsymbol{x}_i^{\mathrm{T}}\boldsymbol{\beta}_0)} x_{i1} \\ \vdots \\ \sum_{i=1}^{n} \frac{\exp(\beta_0 + \boldsymbol{x}_i^{\mathrm{T}}\boldsymbol{\beta}_0)}{1 + \exp(\beta_0 + \boldsymbol{x}_i^{\mathrm{T}}\boldsymbol{\beta}_0)} x_{ip} \end{bmatrix} \\ &= \boldsymbol{X}^{\mathrm{T}}\boldsymbol{y} - \boldsymbol{X}^{\mathrm{T}}\boldsymbol{p}(\boldsymbol{X};\boldsymbol{\beta}) \\ &= \boldsymbol{X}^{\mathrm{T}}[\boldsymbol{y} - \boldsymbol{p}(\boldsymbol{X};\boldsymbol{\beta})],\end{aligned}$$

where $\boldsymbol{p}(\boldsymbol{X};\boldsymbol{\beta}) = [p(\boldsymbol{x}_1;\boldsymbol{\beta}),\ldots,p(\boldsymbol{x}_n;\boldsymbol{\beta})]^{\mathrm{T}}$, the MLE of $\boldsymbol{\beta}$, $\widehat{\boldsymbol{\beta}}$, can be iteratively approximated by using the gradient descent method:

$$\boldsymbol{\beta}_{t+1} = \boldsymbol{\beta}_t + \alpha\boldsymbol{X}^{\mathrm{T}}[\boldsymbol{y} - \boldsymbol{p}(\boldsymbol{X};\boldsymbol{\beta}_t)].$$

For inferential purposes, we have that the asymptotic distribution of $\widehat{\boldsymbol{\beta}}$ is a multivariate normal distribution with vector mean $\boldsymbol{\beta}$ and variance–covariance matrix, the inverse of the negative of the expected value of the Hessian of the log-likelihood, $E\left\{\left[-\frac{\partial \ell(\boldsymbol{\beta};\boldsymbol{y})}{\partial\boldsymbol{\beta}\partial\boldsymbol{\beta}^{\mathrm{T}}}\right]^{-1}\right\} = \left(\boldsymbol{X}^{\mathrm{T}}\boldsymbol{W}\boldsymbol{X}\right)^{-1}$ (McCullagh and Nelder 1989). This is because

$$\begin{aligned}\frac{\partial^2 \ell(\boldsymbol{\beta};\boldsymbol{y})}{\partial\beta_j\partial\beta_k} = \frac{\partial \ell(\boldsymbol{\beta};\boldsymbol{y})}{\partial\beta_j} &= -\sum_{i=1}^{n} \frac{\exp(\beta_0 + \boldsymbol{x}_i^{\mathrm{T}}\boldsymbol{\beta}_0)}{[1 + \exp(\beta_0 + \boldsymbol{x}_i^{\mathrm{T}}\boldsymbol{\beta}_0)]^2} x_{ij}x_{ik} \\ &= -\sum_{i=1}^{n} p(\boldsymbol{x}_i;\boldsymbol{\beta})[1 - p(\boldsymbol{x}_i;\boldsymbol{\beta})]x_{ij}x_{ik}\end{aligned}$$

The Hessian of the log-likelihood is given by

$$\frac{\partial \ell(\boldsymbol{\beta};\boldsymbol{y})}{\partial\boldsymbol{\beta}\partial\boldsymbol{\beta}^{\mathrm{T}}} = \boldsymbol{H} = -\boldsymbol{X}^{\mathrm{T}}\boldsymbol{W}\boldsymbol{X},$$

where $\boldsymbol{W} = \mathrm{Diag}\{p(\boldsymbol{x}_1;\boldsymbol{\beta})[1 - p(\boldsymbol{x}_1;\boldsymbol{\beta})],\ldots,p(\boldsymbol{x}_n;\boldsymbol{\beta})[1 - p(\boldsymbol{x}_n;\boldsymbol{\beta})]\}$.

Once the parameters have been estimated, the prediction response is obtained from the estimated probabilities: $\widehat{y}_o = 1$ if $p\left(\boldsymbol{x}_o;\widehat{\boldsymbol{\beta}}\right) > 0.5$ and $\widehat{y}_o = 0$ if

$p\left(\boldsymbol{x}_o;\widehat{\boldsymbol{\beta}}\right) \leq 0.5$. Of course, a different threshold to 0.5 can be used and this could be considered as a hyperparameter that needs to be tuned in a similar fashion as the penalty parameter in the Ridge procedure.

It is important to point out that the minimization process of the log-likelihood can be performed using a more efficient iterative technique called the Newton–Raphson technique, which is an iterative optimization technique that uses a local quadratic approximation to the log-likelihood function. The following is the Newton–Raphson iterative equation used to search for the beta coefficients:

$$\boldsymbol{\beta}_{t+1} = \boldsymbol{\beta}_t + \boldsymbol{H}^{-1}\boldsymbol{X}^{\mathrm{T}}[\boldsymbol{y} - \boldsymbol{p}(\boldsymbol{X};\boldsymbol{\beta}_t)],$$

where $\boldsymbol{H} = -\boldsymbol{X}^{\mathrm{T}}\boldsymbol{W}\boldsymbol{X}$ is the Hessian matrix whose elements comprise the second derivative of the log-likelihood with regard to the beta coefficients. Therefore, the inverse of the Hessian is $\boldsymbol{H}^{-1} = (\boldsymbol{X}^{\mathrm{T}}\boldsymbol{W}\boldsymbol{X})^{-1}$. So, the previous equation can be expressed as

$$\boldsymbol{\beta}_{t+1} = \boldsymbol{\beta}_t + \left(\boldsymbol{X}^{\mathrm{T}}\boldsymbol{W}\boldsymbol{X}\right)^{-1}\boldsymbol{X}^{\mathrm{T}}[\boldsymbol{y} - \boldsymbol{p}(\boldsymbol{X};\boldsymbol{\beta}_t)].$$

It is important to recall that the Hessian is no longer constant since it depends on $\boldsymbol{\beta}$ through the weighting matrix $\boldsymbol{W}$. Also, it is clear that the logistic regression does not have a closed solution due to the nonlinearity of the logistic sigmoid function. It is important to point out that if instead of maximizing the likelihood we minimize the negative of the log-likelihood, the Newton–Raphson equation for updating the beta coefficients is equal to

$$\boldsymbol{\beta}_{t+1} = \boldsymbol{\beta}_t - \left(\boldsymbol{X}^{\mathrm{T}}\boldsymbol{W}\boldsymbol{X}\right)^{-1}\boldsymbol{X}^{\mathrm{T}}[\boldsymbol{y} - \boldsymbol{p}(\boldsymbol{X};\boldsymbol{\beta}_t)].$$

This Newton–Raphson algorithm for logistic regression is known as the iterative reweighted least squares since the diagonal weighting matrix $\boldsymbol{W}$ is interpreted as variances. Another alternative method for estimating the beta coefficients in logistic regression is the Fisher scoring method that is very similar to the Newton–Raphson method just described, but with the difference that instead of using the Hessian ($\boldsymbol{H}$), it uses the expected value of the Hessian matrix, $E(\boldsymbol{H})$.

3.7.1 Logistic Ridge Regression

Like the MLR, when there is strong collinearity, the variance of the MLE is severely affected and the true effects of the explanatory variables could be falsely identified (Lee and Silvapulle 1988). In a similar fashion as for the MLR, this could be judged directly from the asymptotic covariance matrix of $\widehat{\boldsymbol{\beta}}$. Moreover, in a common prediction context, when the number of features is larger than the number of observations ($p \gg n$), the matrix design is not of full column rank and can cause overfitting, affecting the expected classification error (generalization error) when using the "MLE." One way to avoid overfitting is by replacing the MLE with a regularized MLE as the Ridge MLE estimator of MLR. This is defined as

$$\widetilde{\boldsymbol{\beta}}_s^R(\lambda) = \underset{\boldsymbol{\beta}_s}{\operatorname{argmax}}\left[\ell(\boldsymbol{\beta}_s;\boldsymbol{y}) - \lambda\sum_{j=1}^{p}\beta_{js}^2\right],$$

where λ is a hyperparameter that has a similar interpretation as in the MLR.

In the literature, there are some algorithms that approximate the Ridge estimation. For example, Genkin et al. (2007) used a cyclic coordinate descent optimization algorithm to approximate this. The one-dimensional optimization problem involved is solved by a modified Newton–Raphson method. Another method was proposed by Friedman et al. (2008) in a more general context. Given the current values of $\widetilde{\boldsymbol{\beta}}_s(\lambda)$, the next update of coordinate β_k is given by

$$\beta_{ks} = \frac{\sum_{i=1}^{n} w_i y_{ij}^* x_{ij}}{\sum_{i=1}^{n} w_i x_{ij}^2 + \lambda}$$

with $y_{ij}^* = y_i^* - \widetilde{\mu}(\lambda) - \sum_{\substack{j=1 \\ j\neq k}}^{p} x_{ijs}\widetilde{\beta}_{js}(\lambda)$ for $k = 1, \ldots, p$, and of μ is given by

$$\mu = \frac{\sum_{i=1}^{n} w_i e_i^*}{\sum_{i=1}^{n} w_i}$$

with $e_i^* = y_i^* - \sum_{j=1}^{p} x_{ijs}\widetilde{\beta}_{js}(\lambda)$, where $y_i^* = \widetilde{\beta}_0(\lambda) + \boldsymbol{x}_i^{\mathrm{T}}\widetilde{\boldsymbol{\beta}}_{0s}(\lambda) + \frac{y_i - p\left(\boldsymbol{x}_i;\widetilde{\boldsymbol{\beta}}_s(\lambda)\right)}{w_i}$ and $w_i = p\left(\boldsymbol{x}_i;\widetilde{\boldsymbol{\beta}}_s(\lambda)\right)\left[1 - p\left(\boldsymbol{x}_i;\widetilde{\boldsymbol{\beta}}_s(\lambda)\right)\right]$, $i = 1, \ldots, n$, are pseudo responses and weights that change across the updates. This can be obtained by maximizing, with respect to β_{ks}, the next quadratic approximation of the penalized likelihood at the current values of $\boldsymbol{\beta}_s$ ($\widetilde{\boldsymbol{\beta}}_s(\lambda)$)

$$\ell(\boldsymbol{\beta}_s;\boldsymbol{y}) - \lambda\sum_{j=1}^{p}\left|\beta_{js}\right| \approx \ell^*(\boldsymbol{\beta}_s;\boldsymbol{y}) - \frac{\lambda}{2}\sum_{j=1}^{p}\beta_{js}^2 + c,$$

where $\ell^*(\boldsymbol{\beta}_s;\boldsymbol{y}) = -\frac{1}{2}\sum_{i=1}^{n} w_i\left(y_i^* - \mu - \sum_{j=1}^{p} x_{ijs}\beta_{js}\right)^2$ is the quadratic approximation of $\ell(\boldsymbol{\beta}_s;\boldsymbol{y})$ at current values of $\boldsymbol{\beta}_s$, $\widetilde{\boldsymbol{\beta}}_s(\lambda)$, and c is a constant that does not depend on $\boldsymbol{\beta}_s$. More details of this implementation (glmnet R package) can be found in Friedman et al. (2008). Note that under this approximation, the beta coefficients can be updated by

$$\widehat{\boldsymbol{\beta}}_s^R(\lambda) = \left(\boldsymbol{X}_s^{\mathrm{T}} \boldsymbol{W} \boldsymbol{X}_s + \lambda \boldsymbol{D}\right)^{-1} \boldsymbol{X}_s^{\mathrm{T}} \boldsymbol{W} \boldsymbol{y},$$

where $\boldsymbol{W} = \mathrm{Diag}(w_1, \ldots, w_n)$ and $\boldsymbol{D}$ is the diagonal matrix as defined before.

3.7.2 *Lasso Logistic Regression*

The Lasso penalization can be applied to other models (Tibshirani 1996). In particular, for logistic regression, the Lasso estimator of $\boldsymbol{\beta}_s$ is defined as

$$\widetilde{\boldsymbol{\beta}}_s^L(\lambda) = \underset{\boldsymbol{\beta}_s}{\mathrm{argmax}} \, \ell_L(\boldsymbol{\beta}_s; \boldsymbol{y}),$$

where $\ell_L(\boldsymbol{\beta}_s; \boldsymbol{y}) = \ell(\boldsymbol{\beta}_s; \boldsymbol{y}) - \lambda \sum_{j=1}^{p} |\beta_{js}|$ and is often known as the regularized Lasso likelihood. Numerical methods are also required to obtain this Lasso estimate. There are several possibilities (Genkin et al. 2007), such as non-quadratic programming and iteratively reweighted least squares (Tibshirani 1996), but here we will briefly describe the one proposed by Friedman et al. (2008) and implemented in the glmnent R package. This method consists of applying the coordinate descent procedure to a penalized reweighted least square, which is formed by making a Taylor approximation to the likelihood around the current values of the coefficients. That is, this procedure consists of successively updating the parameters by

$$\widetilde{\boldsymbol{\beta}}_s = \underset{\boldsymbol{\beta}_s}{\mathrm{argmin}} \left(-\ell^*(\boldsymbol{\beta}_s; \boldsymbol{y}) + \frac{\lambda}{2} \sum_{j=1}^{p} |\beta_{js}| \right),$$

where $\ell^*(\boldsymbol{\beta}_s; \boldsymbol{y}) = -\frac{1}{2} \sum_{i=1}^{n} w_i \left(y_i^* - \mu - \sum_{j=1}^{p} x_{ijs} \beta_{js} \right)^2 + c$, $y_i^* = \widetilde{\beta}_0(\lambda) + \boldsymbol{x}_i^{\mathrm{T}} \widetilde{\boldsymbol{\beta}}_{0s}(\lambda) + \frac{y_i - p\left(\boldsymbol{x}_i; \widetilde{\boldsymbol{\beta}}_s(\lambda)\right)}{w_i}$, and $w_i = p\left(\boldsymbol{x}_i; \widetilde{\boldsymbol{\beta}}_s(\lambda)\right)\left[1 - p\left(\boldsymbol{x}_i; \widetilde{\boldsymbol{\beta}}_s(\lambda)\right)\right]$, $i = 1, \ldots, n$, as defined in the Ridge regression case. More details of this implementation can be consulted in Friedman et al. (2008).

Example 5 In this example, we used data corresponding to 40 lines planted with four repetitions. For illustrative purposes, we will use as response a binary variable based on Plant Height. The matrix of features used here was obtained from the genomic relationship ($\boldsymbol{G}$), $\boldsymbol{X} = \boldsymbol{Z}_L \boldsymbol{G}^{1/2}$, where $\boldsymbol{G}^{1/2}$ is the square root matrix of $\boldsymbol{G}$.

The performance of logistic regression, logistic Ridge regression, and Lasso logistic regression for this data set was evaluated across 100 random splittings of the complete data set: 20% for testing (evaluation performance) and 80% for training. The performance was measured by the proportion of cases correctly classified (PCCC) in the testing data. These results are summarized in Table 3.2,

Table 3.2 Performance of the standard, Ridge, and Lasso logistic regression models

Method	PCCC	SD
SLR	0.7284	0.0765
Ridge	0.7240	0.0763
Lasso	0.7206	0.0705

PCCC denotes the proportion of cases correctly classified

where for each method the mean (PCCC) and standard deviation (SD) of the PCCC across the 100 splittings are reported. The table indicates that, on average, the standard logistic (SLR) approach shows slightly better performance than the other two approaches, even better than the Lasso solution. Out of the 100 random partitions, 72, 24, and 4, the SLR, the logistic Ridge regression (LRR), and the logistic lasso regression (LLR) resulted in the higher PCCC value, respectively. However, the difference in the performance of the three methods is not significant because of the large deviation obtained across the different partitions.

The computations were done with the help of the glmnet R package using the following R code:

```
#########################R code for Example 5#######################
load(file ='dat-E3.5.RData')
dat_F = dat$dat_F
dat_F = dat_F[order(dat_F$Rep,dat_F$GID),]
head(dat_F)
G = dat$G
dat_F$y = dat_F$Height
ZL = model.matrix(~0+GID,data=dat_F)
colnames(ZL)
Pos = match(colnames(ZL),paste('GID',colnames(G),sep=''))
max(abs(diff(Pos)))
y = dat_F$y
ei = eigen(G)
X = ZL%*%ei$vectors%*%diag(sqrt(ei$values))%*%t(ei$vectors)
n = length(y)
library(glmnet)
#5FCV
set.seed(1)
Tab = data.frame()
set.seed(1)
for(k in 1:100)
{
 Pos_tr = sample(1:n,n*0.8)
 y_tr = y[Pos_tr] ; X_tr = X[Pos_tr,]; n_tr = dim(X_tr)[1]
 y_tst = y[-Pos_tr]; X_tst = X[-Pos_tr,]
 #Partition for internal training the model
 Grpv_k = findInterval(cut(sample(1:n_tr,n_tr),breaks=5),1:n_tr)
 #RR
 A_RR = cv.glmnet(X_tr,y_tr,family='binomial',
          alpha=0,foldid=Grpv_k,type.measure='class')
 yp_RR = as.numeric(predict(A_RR,newx=X_tst,s='lambda.min',
type='class'))
```

```
 #LR
 A_LR = cv.glmnet(X_tr,y_tr,family='binomial',
         alpha=1,foldid=Grpv_k,type.measure='class')
 yp_LR = as.numeric(predict(A_LR,newx=X_tst,s='lambda.min',
type='class'))
 #SLR
 A_SLR = glmnet(X_tr,y_tr,family='binomial',alpha=0,lambda=0)
 yp_SLR = as.numeric(predict(A_SLR,newx=X_tst,type='class'))


 Tab = rbind(Tab,data.frame(PT=k,PCCC_RR = 1-mean(y_tst!=yp_RR),
             PCCC_LR = 1-mean(y_tst!=yp_LR),
             PCCC_SLR = 1-mean(y_tst!=yp_SLR)))
 cat('k = ', k,'\n')
}
Tab
```

Also, in this R code there are four relevant points:

1. Hundred random partitions were implemented, Pos_tr = sample(1:n,n*0.8), with 80% for training and 20% for testing, and for each training set, an inner K=5 fold CV is performed to tune the λ hyperparameter.
2. Also, here Grpv_k contains the information of the K=5 inner folds to tune the hyperparameter λ.
3. The cv.glmnet function with the following input is used: (a) training set (X_tr, y_tr); (b) with alpha=0 to implement a Ridge regression and alpha=1 to implement a Lasso regression; (c) with foldid=Grpv_k containing training and testing sets to tune the hyperparameter λ; (d) with family='binomial' to implement a logistic regression; and (e) with type.measure='class' as a metric for categorical data to measure the prediction performance of the inner testing set to choose the best value of the hyperparameter λ.
4. The function glmnet with family='binomial' and lambda=0 implements the logistic regression but without penalization.

Appendix 1: R Code for Ridge Regression Used in Example 2

The TR_RR.R script file must contain the following:

```
library(epiR)
source("RR.R")
Tr_RR_f<-function(y_tr,X_tr,K=5,KG=100,KR=1)
{
 n_tr = dim(X_tr)[1]
 X_tr_s = scale2_f(X_tr)
 mu    = mean(y_tr)
 y_tr  = y_tr-mu
 #Inner CV
 lambv = lamb_f(X_tr,K=KG,li=1e-7,ls=1-1e-7)
```

```
iPEv_mean = 0
for(ir in 1:KR)
{
 iPE_mat = matrix(0,nr=length(lambv),nc=K)
 MSEP_mat=iPE_mat
 Grpv = findInterval(cut(sample(1:n_tr,n_tr),breaks=K),1:n_tr)
 for(i in 1:K)
 {
  Pos_itr = which(Grpv!=i)
  X_itr  = X_tr_s[Pos_itr,]
  y_itr  = y_tr[Pos_itr];
  y_itst  = y_tr[-Pos_itr];
  dat_itr = data.frame(y=y_itr,X=X_itr)
  n_itr = dim(X_itr)[1]
  betav_itr = A_RR_f_a_V(y_itr,X_itr,lambv)
  yp_mat   = 0 + (X_tr_s[-Pos_itr,])%*%betav_itr
  iPEv    = colMeans((matrix(y_itst,nr=length(y_itst),
               nc=length(lambv))-yp_mat)**2)
  iPE_mat[,i] =iPEv
 }
 iPEv_mean = (ir-1)/ir*iPEv_mean + rowMeans(iPE_mat)/ir

}

#
plot(log(lambv),iPEv_mean)
Pos_o = which.min(iPEv_mean)
lamb_o = lambv[Pos_o]
A_RR = RR_f(y=y_tr+mu,X=X_tr,lamb=lamb_o,Intercept = TRUE,Std=TRUE)
betav_s = A_RR$betav_s
betav_o = A_RR$betav_o
list(lamb_o = lamb_o, betav_s = betav_s,
   betav_o = betav_o,
   lambv=lambv,iPEv_mean=iPEv_mean,X_tr=X_tr,y_tr =y_tr,Grpv=Grpv)
}

Pred_RR_f<-function(y_tst,X_tst,TR_RR)
{
 #betava_RR = TR_RR$betava_RR
 y_tr = TR_RR$y_tr; X_tr = TR_RR$X_tr
 X  = rbind(X_tr,X_tst)
 betav_s = TR_RR$betav_s
 betav_o = TR_RR$betav_o
 yp_tst = c(cbind(1,X_tst)%*%betav_o)
 plot(y_tst,yp_tst); abline(a=0,b=1)
 MSEP_RR = mean((y_tst-yp_tst)**2)
 list(MSEP=MSEP_RR, betav_s = betav_s,betav_o = betav_o)
}
Pred_ols_f<-function(y_tst,X_tst,y_tr,X_tr)
{
 #OLS
 p = dim(X_tr)[2]
 A = lm_f(y_tr,X_tr)
```

```
 sdv  = apply(X_tr,2,sd2_f)
 A_inv = diag(c(1,1/sdv))
 A_inv[1,1] = 1
 A_inv[1,2:(p+1)] = -apply(X_tr,2,mean)/sdv
 betav_s = A$betav_s
 betav_o = A_inv%*%betav_s
 yp_tst_ols = c(cbind(1,X_tst)%*%betav_o)
 MSEP_ols = mean((y_tst-yp_tst_ols)**2)
 list(MSEP=MSEP_ols,betav_s = betav_s)
}
```

The script file accessed by source("RR.R") must contain the following R code:

```
sd2_f<-function(x)
{
 (mean((x-mean(x))**2))^0.5
}
scale2_f<-function(X)
{
 scale(X,center=TRUE,scale = apply(X,2,sd2_f))
}

#RR internal standarized: zij = (xij-\bar{x}_j)/ss_j
#ss_j = mean (xij-\bar{x}_j)^2
RR_f<-function(y,X,lamb = 0, Intercept=TRUE,Std=TRUE)
{
 p   = dim(X)[2]; n = dim(X)[1]
 if(Std == TRUE)
 {
  sdv  = apply(X,2,sd2_f)
  A_inv = diag(c(1,1/sdv))
  A_inv[1,1] = 1
  A_inv[1,2:(p+1)] = -apply(X,2,mean)/sdv
  mu = mean(y)
  X_s  = scale2_f(X)
  Xa = X_s
  svd_X = svd(Xa); d1  = svd_X$d
  Gama1 = svd_X$v
  U = svd_X$u
  pa   = dim(Gama1)[2]
  betav_s = c(mu,Gama1%*%(((d1/(d1^2+lamb)))*(t(U)%*%(y-mu))))
  betav_o = A_inv%*%betav_s# betav in orginal scale
  list(betav_s = betav_s,betav_o=betav_o)
 }
 else
 {
  Xa = X
  sdv  = apply(X,2,sd2_f)
  p = dim(X)[2]
  svd_X = svd(Xa)
  d1  = svd_X$d
  Gama1 = svd_X$v; U = svd_X$u
  betav = Gama1%*%((d1/(d1^2+lamb))*(t(U)%*%(y)))#-mean(y)
```

```
  betav
  #tX = t(X)
  #betav = ginv(tX%*%X+lamb*diag(p))%*%tX%*%y
  #betav
 }
}

RR_f_V = Vectorize(RR_f,'lamb')

A_RR_f_a <-function(y_itr,X_itr,lamb)
{
 A = RR_f(y=y_itr,X=X_itr,lamb=lamb,Std=FALSE,Intercept = FALSE)
 A
}
A_RR_f_a_V<-Vectorize(A_RR_f_a,'lamb')

lamb_f<-function(X,K=100,li=0.001,ls=0.999)
{
 Xac = scale2_f(X)
 n = dim(Xac)[1]
 R2v = seq(li,ls,length=K)
 lambv = (1-R2v)/R2v*sum(diag(Xac%*%t(Xac)))/n
 lambv = exp(seq(min(log(lambv)),max(log(lambv)),length=K))
 sort(lambv,decreasing = TRUE)
}

library(MASS)
lm_f<-function(y,X)
{
 p = dim(X)[2]
 sdv  = apply(X,2,sd2_f)
 A_inv = diag(c(1,1/sdv))
 A_inv[1,1] = 1
 A_inv[1,2:(p+1)] = -apply(X,2,mean)/sdv
 X = scale2_f(X); mu = mean(y);
 svd_X = svd(X); d = svd_X$d
 d1  = svd_X$d; Gama1 = svd_X$v; U = svd_X$u
 betav_s = c(mu,Gama1%*%((1/d1)*(t(U)%*%(y-mu))))
 betav_o = A_inv%*%betav_s# betav in orginal scale
 list(betav_s = betav_s,betav_o=betav_o)
}
```

References

Allaire JJ, Chollet F (2019) keras: R Interface to 'Keras'. R package version, 2(4)

Beysolow T II (2017) Introduction to deep learning using R: a step-by-step guide to learning and implementing deep learning models using R. Apress, San Francisco, CA

Burden RL, Faires JD (2011) Numerical analysis. Cengage Learning, Boston, MA

Casella G, Berger RL (2002) Statistical inference. Duxbury, Thomson Learning, Pacific Grove, CA

Christensen P (2011) Plane answers to complex questions: the theory of linear models. Springer Science+Business Media, New York

Efron B, Hastie T, Johnstone I, Tibshirani R (2004) Least angle regression. Ann Stat 32(2):407–499

Friedman J, Hastie T, Hoeing H, Tibshirani R (2007) Pathwise coordinate optimization. Ann Appl Stat 2(1):302–332

Friedman J, Hastie T, Tibshirani R (2008) Sparse inverse covariance estimation with the graphical lasso. Biostatistics 9(3):432–441

Friedman J, Hastie T, Tibshirani R (2010) Regularization paths for generalized linear models via coordinate descent. J Stat Softw 33(1):1–22

Genkin A, Lewis D, Madigan D (2007) Large-scale Bayesian logistic regression for text categorization. Technometrics 49(3):291–304

Goodfellow I, Bengio Y, Courville A (2016) Deep learning. The MIT Press, Cambridge, MA

Hastie T, Tibshirani R, Friedman J (2009) The elements of statistical learning: data mining, inference, and prediction. Springer, New York

Hastie T, Tibshirani R, Wainwright M (2015) Statistical learning with sparsity: the lasso and generalizations. Chapman and Hall/CRC, New York

Haykin S (2009) Neural networks and learning machines. Pearson Prentice Hall, New Jersey, p 909

James G, Witten D, Hastie T, Tibshirani R (2013) An introduction to statistical learning: with applications in R. Springer Science+Business Media, New York

Lee AH, Silvapulle MJ (1988) Ridge estimation in logistic regression. Commun Stat Simul Comput 178(4):1231–1257. https://doi.org/10.1080/03610918808812723

McCullagh P, Nelder JA (1989) Generalized linear models. Chapman and Hall, London, England

Montgomery DC, Peck EA, Vining GG (2012) Introduction to linear regression. Wiley, Hoboken, NJ

Nocedal J, Wright SJ (2006) Numerical optimization. Springer, New York

Rencher AC, Schaalje GB (2008) Linear models in statistics. Wiley, Hoboken, NJ

Tibshirani R (1996) Regression shrinkage and selection via the lasso. J R Stat Soc B 58(1):267–288

Wakefield J (2013) Bayesian and frequentist regression methods. Springer Science+Business Media, New York

Warner B, Misra M (1996) Understanding neural networks as statistical tools. Am Stat 50 (4):284–293

4

Overfitted and Underfitted Models

4.1 The Problem of Overfitting and Underfitting

The *overfitting* phenomenon occurs when the statistical machine learning model learns the training data set so well that it performs poorly on unseen data sets. In other words, this means that the predicted values match the true observed values in the training data set too well, causing what is known as overfitting. Overfitting happens when a statistical machine learning model learns the systematic and noise (random fluctuations) parts in the training data to the extent that it negatively impacts the performance of the statistical machine learning model on new data. This means that the statistical machine learning model adapts very well to the noise as well as to the signal that is present in the training data. The problem is that these concepts do not apply to independent (new) data and negatively affect the model's ability to generalize. Overfitting is more probable when learning a loss function from a complex statistical machine learning model (with more flexibility). For this reason, many nonparametric statistical machine learning models also include constraints in the loss function to improve the learning process of the statistical machine learning models. For example, artificial neural networks (ANN), mentioned later, are a nonparametric statistical machine learning model that is very flexible and is subject to overfitting training data. This problem can be addressed by dropping out (setting to zero) the weights of a certain percentage of hidden units in order to avoid overfitting.

On the other hand, an *underfitted* phenomenon occurs when few predictors are included in the statistical machine learning model, i.e., it is a very simple model that poorly represents the complete picture of the predominant data pattern. This problem also arises when the training data set is too small or not representative of the population data. An *underfitted* model does a poor job of fitting the training data and for this reason it is not expected to satisfactorily predict new data points. This implies that the predictions using unseen data are weak, since individuals are perceived as strangers unfamiliar with the training data set.

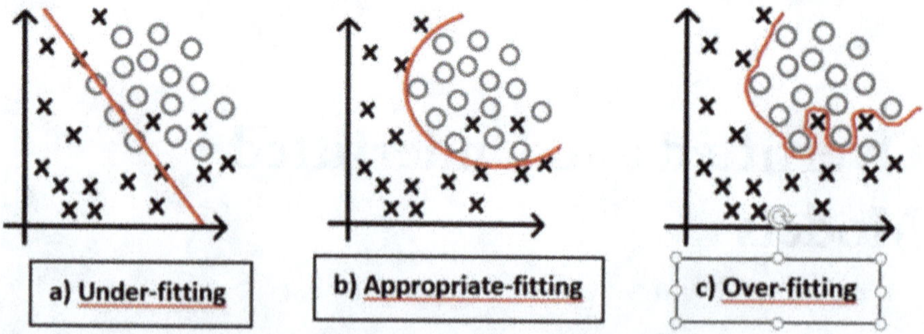

Fig. 4.1 Schematic illustration of three models for classification: (**a**) M1 with underfitting, (**b**) M2 with appropriate fitting, and (**c**) M3 with overfitting

Consider a scattered series of points (y_1, x_1), . . ., (y_n, x_n), on a plane, to which we want to adjust a statistical machine learning method. This means that we are looking for the best $f(\boldsymbol{x}_i)$ that explains the existing relationship between the response variable (y_i) and the predictors ($x_1, \ldots, x_n$). We assume that we have three options for $f(\boldsymbol{x}_i)$: M1, the simple model plotted in Fig. 4.1a; M2, an intermediate model shown in Fig. 4.1b; and M3, a complex model shown in Fig. 4.1c.

Under the classification framework, the first panel in Fig. 4.1 (left side, panel a) shows an unsatisfactory fit (underfitted) since the line does not cover most of the points (has high bias) in the plot. As such, we expect that the prediction of unseen data of this model, M1, will perform badly. In contrast, panel c of Fig. 4.1 shows an almost perfect fit, since the predicted line covers all the data points. While at first glance, you may think that model M3 will perform well when predicting unseen data, this is actually untrue since the predicted line covers all points that are noise and those that are signal (overfit); for this reason, this type of model also performs poorly in the prediction of future data due to its complexity and high variance. Therefore, the best option for predicting unseen data is model M2 (Fig. 4.1, panel b) since it represents the predominant (smooth) pattern enough to represent the apparent data pattern while maintaining a balance between bias and variance. For this reason, a well-fitted model is one that faithfully represents the sought-after predominant pattern in the data, while ignoring the idiosyncrasies in the training data. As such, a well-fitted model in the testing set should be in the neighborhood of the model's accuracy based on the training data set, that is, the model's accuracy in the testing set should be approximately equal to that of the model's accuracy in the training set. In contrast, an overfitted model in the testing data set will be far from the neighborhood of the model's accuracy based on the training data set, and usually its prediction performance is very high (good) in the training set and consequently low (bad) in the testing set (Ratner 2017).

The paradox of overfitting is defined as complex models that contain more information about the training data, but less information about the testing data (future data we want to predict). In statistical machine learning, overfitting is a major issue and leads to some serious problems in research: (a) some relationships that seem statistically significant are only noise, (b) the complexity of the statistical machine learning model is very large for the amount of data provided, and (c) the model in general is not replicable and predicts poorly.

Since the main goal of developing and implementing statistical machine learning methods is to predict unseen data not used for training the statistical machine learning algorithm, researchers are mainly interested in minimizing the testing error (generalization error applicable to future samples) instead of minimizing the training error that is applicable to the observed data used for training the statistical machine learning algorithm.

According to Shalev-Shwartz and Ben-David (2014), if the learning fails, these are some approaches to follow:

1. Increase the sample size of the training set.
2. Modify the hypothesis by (a) enlarging it, (b) reducing it, (c) completely changing it, and (d) changing the parameters being used. We understand a hypothesis as the models and their parameters under evaluation. This point is very important to reach a reasonable model for your data.
3. Change the feature representation of the data.
4. Change the statistical machine learning algorithm used.

4.2 The Trade-Off Between Prediction Accuracy and Model Interpretability

Accuracy is the ability of a statistical machine learning model to make correct predictions and those models with more complexity (called flexible models) are better in terms of accuracy, while the simple, less complex models (called inflexible models) are less accurate but more interpretable. Interpretability indicates to what degree the model allows for human understanding of natural phenomena. For these reasons, when the goal of the study is prediction, flexible models should be used; however, when the goal of the study is inference, inflexible models are more appropriate because they more easily interpret the relationship between the response variables and the predictor variables. As the complexity of the statistical machine learning model increases, the bias is reduced and the variance increases. For this reason, when more parameters are included in the statistical machine learning model, the complexity of the model increases and the variance becomes the main concern while the bias steadily falls. For example, James et al. (2013) state that the linear regression model is a relatively inflexible method because it only generates linear functions, while the support vector machine method is one of the most flexible statistical machine learning methods.

Before providing an analytical interpretation of the trade-off between the bias and variance, we must understand the meaning of both concepts. Bias is the difference between the expected prediction of our statistical machine learning model and the true observed values. For example, assume that you poll a specific city where half of

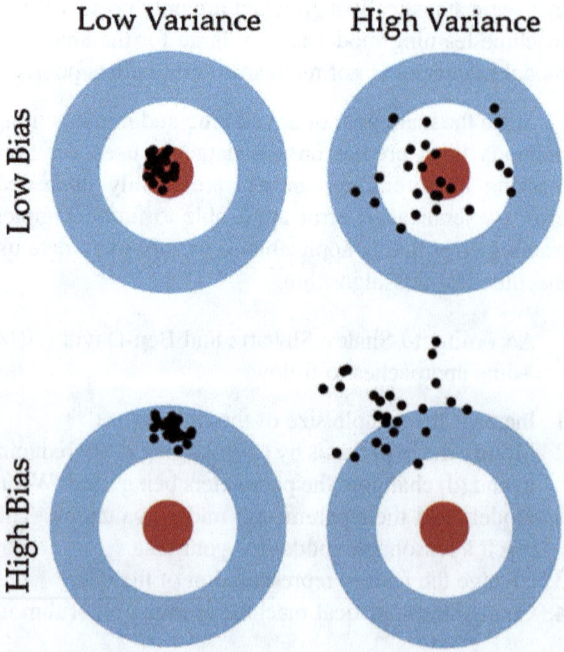

Fig. 4.2 Graphical representations of different levels of bias and variance

the population is high-income and the other half is low-income. If you collected a sample of high-income people, you would conclude that the entire city has high income. This means that your conclusion is heavily biased since you only sampled people with high income. On the other hand, error variance refers to the amount that the estimate of the objective function will change using a different training data set. In other words, the error variance accounts for the deviation of predictions from one repetition to another using the same training set. Ideally, when a statistical machine learning model with low error variance predicts a value, the predicted value should remain almost the same, even when changing from one training data set to another; however, if the model has high variance, then the predicted values of the statistical machine learning method are affected by the values of the data set. We provide a graphical visualization of bias and variance with a bull's eye diagram (see Fig. 4.2). We assume that the center of the target is a statistical machine learning model that perfectly predicts the correct answers. As we move away from the bull's eye, our predictions get worse. Let us assume that we can repeat our entire statistical machine learning model building process to get a number of separate hits on the target. Each hit represents an individual realization of our statistical machine learning model, given the chance variability in the training data we gathered. Sometimes we accurately predict the observations of interest since we captured a representative sample in our training data, while other times we obtain unreliable predictions since our training data may be full of outliers or nonrepresentative values. The four combinations of cases resulting from both high and low bias and variance are shown in Fig. 4.2.

Burger (2018) concludes that the best scenario is the one with *low bias* and *low variance*, since samples are acceptable representatives of the population (Fig. 4.2), while in the case of *high bias* and *low variance*, the samples are fairly consistent, but not particularly representative of the population (Fig. 4.2). However, when there is *low bias* and *high variance*, the samples vary widely in their consistency, and only some may be representative of the population (Fig. 4.2). Finally, with *high bias* and *high variance*, the samples are somewhat consistent, but unlikely to be representative of the population (Fig. 4.2).

If we denote the variable we are trying to predict as y and our covariates as $\boldsymbol{x}_i$, we may assume that there is a relationship between y and $\boldsymbol{x}_i$, as that given in Eq. (1.1) from Chap. 1, where the error term is normally distributed with a mean of zero and variance σ^2. The expected prediction error for a new observation with value x, using a quadratic loss function, is given by Hastie et al. (2008, page 223):

$$\begin{aligned} E\left(y-\widehat{f}\{\boldsymbol{x}_i\}\right)^2 &= E(y-f\{\boldsymbol{x}_i\})^2+\left(E\left(\widehat{f}\{\boldsymbol{x}_i\}\right)-f\{\boldsymbol{x}_i\}\right)^2+E\left\{\widehat{f}(\boldsymbol{x}_i)-E\left(\widehat{f}\{\boldsymbol{x}_i\}\right)\right\}^2 \\ &= \mathrm{Var}(y)+\left[\mathrm{Bias}\left(\widehat{f}\{\boldsymbol{x}_i\}\right)\right]^2+\mathrm{Var}\left(\widehat{f}(\boldsymbol{x}_i)\right) \\ &= \mathrm{Var}(\epsilon)+\left[\mathrm{Bias}\left(\widehat{f}\{\boldsymbol{x}_i\}\right)\right]^2+\mathrm{Var}\left(\widehat{f}(\boldsymbol{x}_i)\right), \end{aligned}$$

where Bias is the result of misspecifying the statistical model f. Estimation variance (the third term) is the result of using a sample to estimate f. The first term is the error (irreducible error) that results even if the model is correctly specified and accurately estimated. This irreducible error is the noise term in the true relationship that cannot fundamentally be reduced by any model. Given the true model and infinite data to train (calibrate) it, we should be able to reduce both the bias and variance terms to 0. However, in a world with imperfect models and finite data, there is a trade-off between minimizing the bias and minimizing the variance. The above decomposition reveals a source of the difference between explanatory and predictive modeling: In explanatory modeling, the focus is on minimizing bias to obtain the most accurate representation of the underlying theory. In contrast, predictive modeling seeks to minimize the combination of bias and estimation variance, occasionally sacrificing theoretical accuracy for improved empirical precision (Shmueli 2010).

These four aspects impact every step of the modeling process, such that the resulting f is markedly different in the explanatory and predictive contexts.

Let us assume that f is a reasonable operationalization of the true function (F) relating constructs X and Y. Choosing a function f^* that is intentionally biased in place of f is very undesirable from a theoretical–explanatory standpoint. However, the election of f^* is desirable to f under the prediction approach. We show this using the statistical model $y = \beta_1 x_1 + \beta_2 x_2 + \beta_3 x_3 + \epsilon$, which is assumed to be correctly specified with respect to F. Using data, we obtain the estimated model $\widehat{f}$, which has the following properties:

$$\text{Bias} = 0$$

$$\text{Var}\left(\widehat{f}(\boldsymbol{x}_i)\right) = \text{Var}\left(\widehat{\beta}_1 x_1 + \widehat{\beta}_2 x_2 + \widehat{\beta}_3 x_3\right) = \sigma^2 \boldsymbol{x}^{\text{T}} \left(\boldsymbol{X}\boldsymbol{X}^{\text{T}}\right)^{-1} \boldsymbol{x},$$

where $\boldsymbol{x}$ is the vector $\boldsymbol{x} = [x_1, x_2, x_3]^{\text{T}}$ and $\boldsymbol{X}$ is the design matrix based on all predictors. Combining the squared bias with the variance gives, as expected, the prediction error (EPE).

$$E\left(y - \widehat{f}\{\boldsymbol{x}_i\}\right)^2 = \sigma^2 + 0 + \sigma^2 \boldsymbol{x}^{\text{T}} \left(\boldsymbol{X}\boldsymbol{X}^{\text{T}}\right)^{-1} \boldsymbol{x} = \sigma^2 \left[1 + \boldsymbol{x}^{\text{T}} \left(\boldsymbol{X}\boldsymbol{X}^{\text{T}}\right)^{-1} \boldsymbol{x}\right].$$

In comparison, consider the estimated underspecified form $\widehat{f}^*(\boldsymbol{x}_i) = x_1 \widehat{\gamma}$. The bias and variance here are

$$\text{Bias} = E\left(\widehat{f}\{\boldsymbol{x}_i\}\right) - f\{\boldsymbol{x}_i\} = x_1 \left(x_1 x_1^{\text{T}}\right)^{-1} x_1^{\text{T}} (\beta_1 x_1 + \beta_2 x_2 + \beta_3 x_3) - (\beta_1 x_1 + \beta_2 x_2 + \beta_3 x_3)$$

$$\text{Var}\left(\widehat{f}(\boldsymbol{x}_i)\right) = \sigma^2 x_1 \left(x_1 x_1^{\text{T}}\right)^{-1} x_1^{\text{T}}$$

Combining the squared bias with the variance EPE is equal to

$$E\left(y - \widehat{f}\{\boldsymbol{x}_i\}\right)^2 = \left[x_1 \left(x_1 x_1^{\text{T}}\right)^{-1} x_1^{\text{T}} (x_2 \beta_2 + x_3 \beta_3) - (x_2 \beta_2 + x_3 \beta_3)\right]^2 + \sigma^2 \left[1 + x_1 \left(x_1 x_1^{\text{T}}\right)^{-1} x_1^{\text{T}}\right].$$

Although the bias of the underspecified model $f^*(\boldsymbol{x}_i)$ is larger than that of $f\{\boldsymbol{x}_i\}$, its variance can be smaller, and in some cases, so small that the overall EPE will be lower for the underspecified model. Wu et al. (2007) showed the general result for an underspecified linear regression model with multiple predictors. In particular, they showed that the underspecified model that leaves out q predictors has a lower EPE when the following inequality holds:

$$q\sigma^2 > \beta_2^{\text{T}} X_2^{\text{T}} (I - H_1) X_2 \beta_2$$

This means that the underspecified model produces more accurate predictions, in terms of lower EPE, in the following situations: (a) when the data are very noisy (large σ^2); (b) when the true absolute values of the excluded parameters (in our example, β_2 and β_3) are small; (c) when the predictors are highly correlated; and (d) when the sample size is small or the range of left-out variables is small (Shmueli 2010).

Hagerty and Srinivasan (1991) nicely summarize this situation: "We note that the practice in applied research of concluding that a model with a higher predictive validity is "truer," is not a valid inference. This paper shows that a parsimonious but less true model can have a higher predictive validity than a truer but less parsimonious model."

4.3 Cross-validation

Cross-validation (CV) is a strategy for model selection or algorithm selection. CV consists of splitting the data (at least once) for estimating the error of each algorithm. Part of the data (the training set) is used for training each algorithm, and the remaining part (the testing set) is used for estimating the error of the algorithm. Then, CV selects the algorithm with the smallest estimated error. For this reason, CV is used to evaluate the prediction performance of a statistical machine learning model in out-of-sample data. This technique ensures that the data used for training the statistical machine learning model are independent of the testing data set in which the prediction performance is evaluated. It consists of repeating and recording the arithmetic average obtained from the evaluation measures on different partitions. Under k-fold CV, which is explained in greater detail later, this process is repeated a total of k times, with each of the k groups getting the chance to play the role of the test data, and the remaining $k - 1$ groups used as training data. In this way, we obtain k different estimates of the prediction error. As prediction performance is reported the average of these estimates of prediction error. CV is used in data analysis to validate the implemented models where the main objective is prediction and to estimate the prediction performance of a statistical learning model that will be carried out in practice. In other words, CV evaluates how well the statistical machine learning model generalized new data not used for training the model. The results of the CV largely depend on how the division between the training and testing sets is carried out. For this reason, in the following sections, we provide the more popular types of CV used in the implementation of statistical learning models.

4.3.1 The Single Hold-Out Set Approach

The single hold-out set or validation set approach consists of randomly dividing the available data set into a training set and a validation or hold-out set (Fig. 4.3). The statistical machine learning model is trained with the training set while the hold-out

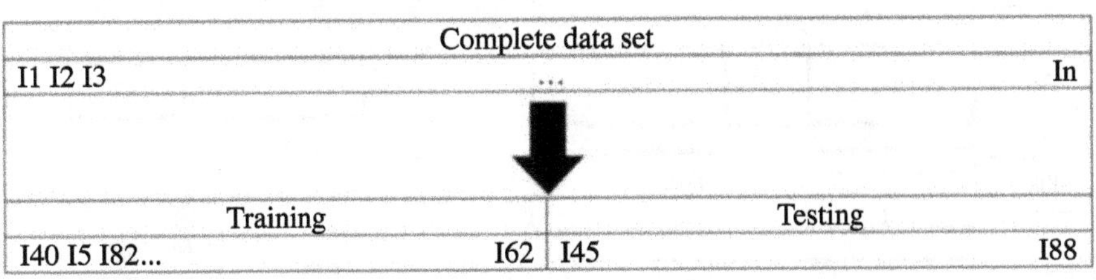

Fig. 4.3 Schematic representation of the hold-out set approach. A set of observations are randomly split into a training set with individuals I40, I5, I82, among others, and into a testing set with observations I45, I88, among others. The statistical machine learning model is fitted on the training set and its performance is evaluated on the validation set (James et al. 2013)

set (testing set) is used to study how well that statistical machine learning model performs on unseen data. For example, 80% of the data can be used for training the model and the remaining 20% of the data for testing it. One weakness of the hold-out (validation) set approach is that it depends on just one training-testing split and its performance depends on how the data are split into the training and testing sets.

4.3.2 The k-Fold Cross-validation

In k-fold CV, the data set is randomly divided into k complementary folds (groups) of approximately equal size. One of the subsets is used as testing data and the rest ($k-1$) as training data. Then $k-1$ folds are used for training the statistical machine learning model and the remaining fold for evaluating the out-of-sample prediction performance. For these reasons, the statistical machine learning model is fitted k times using a different partition (fold) as the testing set and the remaining $k-1$ as the training set. Finally, the arithmetic mean of the k folds is obtained and reported as the prediction performance of the statistical machine learning model (see Fig. 4.4). This method is very accurate because it combines k measures of fitness resulting from the k training and testing data sets into which the original data set was divided, but at the cost of more computational resources. In practice, the choice of the number of folds depends on the measurement of the data set, although 5 or 10 folds are the most common choices.

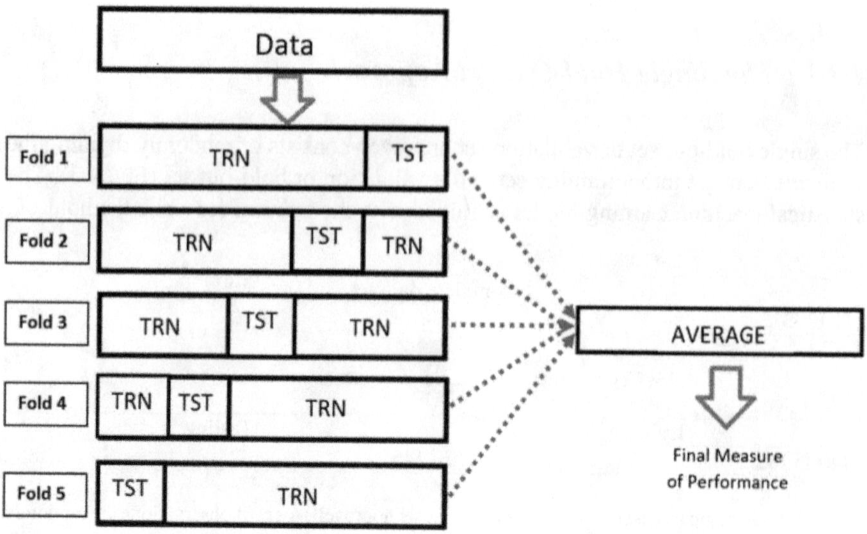

Fig. 4.4 Schematic representation of the k-fold cross-validation with complementary subsets with $k = 5$

It is important to point out that to reduce variability, we recommend implementing the k-fold CV multiple times, each time using different complementary subsets to form the folds; the validation results are combined (e.g., averaged) over the rounds (times) to give a better estimate of the statistical machine learning model predictive performance.

4.3.3 The Leave-One-Out Cross-validation

The Leave-One-Out (or LOO) CV is very simple since each training data set is created by including all the individuals except one, while the testing set only includes the excluded individual. Thus, for n individuals in the full data set, we have n different training and testing sets. This CV scheme wastes minimal data, as only one individual is removed from the training set.

Regarding the k-fold cross-validation that was just explained, n models are built from n individuals (samples) instead of k models, where $n > k$. Moreover, each model is trained on $n - 1$ samples rather than $(k - 1)n/k$. In both cases, since k is normally not too large and $k < n$, LOO is more computationally expensive than k-fold cross-validation. In terms of prediction performance, LOO normally produces high variance for the estimation of the test error. Because $n - 1$ of the n samples is used to build each statistical machine learning model, those constructed from folds are virtually identical to each other and to the model built from the entire training set. However, when the learning curve is steep for the evaluated training size, then five- or ten-fold cross-validation usually overestimates the generalization error.

Learning curves (LC) are considered effective tools to monitor the performance of the employee exposed to a new task. LCs provide a mathematical representation of the learning process that takes place as the task is repeated. In statistical machine learning the LC is a line plot of learning (y-axis) over experience (x-axis). Learning curves are extensively used in statistical machine learning for algorithms that learn (their parameters) incrementally over time, such as deep neural networks. In general, there is considerable empirical evidence suggesting that five- or ten-fold cross-validation should be preferred to LOO.

4.3.4 The Leave-m-Out Cross-validation

The Leave-m-Out (LmO) CV is very similar to LOO as it creates all the possible training/test sets by removing m samples from the complete set. For n samples, this produces $\binom{n}{m}$ train-test pairs. Unlike LOO and k-fold, the test sets will overlap for $m > 1$. For example, in a Leave-2-Out CV with a data set with four samples (I1, I2,

I3, and I4), the total number of training-testing sets is equal to $\binom{4}{2} = 6$; this means that the six testing sets are: [I3, I4] [I1, I2], [I2, I4] [I1, I3], [I2, I3] [I1, I4], [I1, I4] [I2, I3],[I1, I3] [I2, I4], and [I1,I2] [I3, I4], while the training sets are the complementary elements of each testing set.

4.3.5 Random Cross-validation

In this type of CV, the number of partitions (independent training-testing data set splits) is defined by the user, and more partitions are better. Each partition is generated by randomly dividing the whole data set into two subsets: the training (TRN) data set and the testing (TST) data set. The percentage of the whole data set assigned to the TRN and TST data sets is also fixed by the user. For example, for each random partition, the user can decide that 80% of the whole data set can be assigned to the TRN data set and the remaining 20% to the TST data set. Random cross-validation is different from *k*-fold cross-validation because the partitions are not mutually exclusive; this means that in the random cross-validation approach, one observation can appear in more than one partition. Consequently, some samples cannot be evaluated, whereas others can be evaluated more than once, meaning that the testing and training subsets can be superimposed (Montesinos-López et al. 2018a, b). To control the randomness for reproducibility, we recommend using a specific seed in the random number generator. We recommend using at least ten random partitions to obtain enough accuracy in the estimate of prediction performance.

4.3.6 The Leave-One-Group-Out Cross-validation

The Leave-One-Group-Out (LOGO) CV is useful when individuals are grouped (in environments or years, or even another criterion), where the number of groups (g) is at least two and the information of $g - 1$ groups are used as the training set while all individuals of the remaining group are used as the testing set. For example, in the context of genomic selection, when the plant breeder is interested in predicting these lines in another environment, the same (or different) lines were frequently evaluated in g environments or years, that represent the groups. Jarquín et al. (2017) denotes this type of CV strategy as CV1 in the context of plant breeding. Under this approach, the predictions are reported for each of the g groups because the scientist is interested in the prediction performance of each environment. Many times, the groups are the years under study and the aim is to predict the information of a complete year. However, when the groups are years and if we suspect that there is a considerable correlation between observations that are near in time. Therefore, it is

<table>
<tr><td>Fold 1</td><td>TRN</td><td>TST</td><td></td><td></td><td></td></tr>
<tr><td>Fold 2</td><td colspan="2">TRN</td><td>TST</td><td></td><td></td></tr>
<tr><td>Fold 3</td><td colspan="3">TRN</td><td>TST</td><td></td></tr>
<tr><td>Fold 4</td><td colspan="4">TRN</td><td>TST</td></tr>
<tr><td></td><td>Year 1</td><td>Year 2</td><td>Year 3</td><td>Year 4</td><td>Year 5</td></tr>
</table>

Fig. 4.5 Schematic representation of time series data with 5 years, using the previous year for predicting the next year. However, in some practical applications, we are not interested in going too far back in time since the training and testing sets will be less related the farther you go

imperative to evaluate our statistical machine learning model for time series data on "future" observations. In this sense, the training sets are composed of the previous years to predict the subsequent year. This method can also be seen as a variation of k-fold CV, where the first folds are used for training the statistical machine learning model and the fold $(k + 1)$ is the corresponding testing set. The main difference in this CV method is that successive training sets are supersets of those that come before them. Also, it adds all surplus data to the first training partition, which is always used to train the model (see Fig. 4.5).

Figure 4.5 shows that under this type of CV, for time series data, the predictions are for $g - 1$ years; individuals from the first year are not predicted since a training set is not available.

4.3.7 Bootstrap Cross-validation

First, we will define the bootstrapping method to understand how it is used in the CV approach, which should then be straightforward. Bootstrapping is a type of resampling method where, for example, $B = 10$ samples of the same size are repeatedly drawn, with replacement, from a single original sample. Afterward, each of these B samples is used to estimate statistics (for example, the mean, variance, median, minimum, etc.) of a population, and the average of all the B sample estimates of the target statistic is reported as the final estimate. In the context of statistical machine learning, these samples are used to evaluate the prediction performance of the algorithm under study for unseen data. One important difference between this CV approach and all the procedures explained above is that now the training set has the same size (number of observations) as the original sample because the bootstrap method replaced some individuals more than once. According to Kuhn and Johnson (2013), as a result, some observations will be represented multiple times in the bootstrap sample, while others will not be selected at all; those observations not selected are referred to as the testing set, however, this CV strategy is quite different than the previously explained. Efron (1983) pointed out that the prediction performance of the bootstrap samples tends to have less uncertainty than the k-fold cross-validation since on average, 63.2% of the data points are represented (for training) at least once in any sample size. For this reason,

Fold	Training	Testing
1	I7, I4, I6, I9, I2, I3, I4, I4, I8, I6, I8, I7	I1, I5, I10, I11, I12
2	I2, I8, I5, I6, I1, I4, I5, I11, I1 1,I8,I10, I5	I3, I7, I9, I12
3	I5, I9, I11, I3, I10, I5, I7, I2, I 3, I11, I6, I9	I1, I4, I8, I12
4	I10, I12, I9, I7, I4, I3, I1, I9, I3, I2, I1, I6	I5, I8, I11
5	I2, I10, I5, I12, I3, I6, I3, I7, I6, I6, I5, I7	I1, I4, I8, I9

Fig. 4.6 Schematic representation of bootstrap cross-validation

this CV approach has a bias similar to implementing a $k =$ two fold cross-validation, and as the training set becomes smaller, the bias becomes more problematic. To understand this CV method, we provide a simple example of how the training and testing samples are constructed. If we have a sample with 12 individuals denoted as I1, I2, ..., I12, we will select $B = 5$ bootstrap samples. Each bootstrap sample is obtained with replacement and the individuals that appear in each one correspond to the training sample; those that are not present will correspond to the testing set. Figure 4.6 provides the five bootstrap samples; each training sample has the same size as the original, however, only some individuals appear in each bootstrap sample, while those individuals that do not appear are included in the testing set. For example, in the first fold, the training bootstrap sample contains seven different individuals (I2, I3, I4, I6, I7, I8, and I9), while the testing set contains five individuals (I1, I5, I10, I11, and I12). It is important to point out that since the training sample has the same size as the original sample, some individuals in the training sample are repeated at least twice; in the first fold, the individual I4 are repeated three times, whereas I6, I7, and I8 are repeated twice. Finally, similar to other methods, the statistical machine learning model is trained with each training set; likewise, the prediction performance of the model is evaluated in each testing set. The average of these sample predictions is reported as the estimated testing error.

4.3.8 *Incomplete Block Cross-validation*

Incomplete block (IB) CV should be used when there are J treatments evaluated in I blocks and the same treatments are evaluated in all the blocks. The idea behind this

Table 4.1 TRN data set for $I = 3$ blocks and $J = 10$ treatments with 70% of data for training

Block	Treatments per block						
1	1	3	5	7	8	9	10
2	2	3	4	5	6	7	9
3	1	2	4	6	7	8	10

CV method is that some treatments should be present in some blocks but absent in others, whereas the same treatment should be present in at least one environment (block). The theory of incomplete block designs developed in the experimental designs statistical area can be used to construct the training set. For example, under a balanced incomplete block (BIB) design, the term incomplete means that all treatments in each block cannot be evaluated, whereas balanced means that each pair of treatments occur together λ times. The training set is constructed by first defining the % of individuals in the TRN set using the equation $sI = Jr = N_{TRN}$, where J represents the number of treatments under study, I represents the number of blocks under study, r denotes the number of repetitions of each treatment, and s denotes the treatments per block. For example, suppose that we had $J = 10$ treatments and $I = 3$ blocks (that is, 30 individuals), and we decided to use $N_{TRN} = 21$ (70%) of the total individuals in the TRN set. Therefore, the number of treatments by block can be obtained by solving ($sI = N_{TRN}$) for s, which results in $s = N_{TRN}/I$. This means that $s = 21/3 = 7$ treatments per block. Then, the corresponding elements for the training set can be obtained with the function find.BIB(10, 3, 7) of the package crossdes of the R statistical software. The numbers used in the function find.BIB() denote the treatments, the blocks, and the treatments per block, respectively. Finally, the treatments that make up the TRN set are shown in Table 4.1.

According to Table 4.1, it is clear that each treatment is present in two blocks and missing in one block. For example, in Block 1 the testing set includes treatments 2, 4, and 6; in Block 2, the testing set is composed of treatments 1, 8, and 10; and in Block 3, the testing set is composed of treatments 3, 5, and 9.

4.3.9 Random Cross-validation with Blocks

Random cross-validation with blocks was proposed by Lopez-Cruz et al. (2015) and belongs to the so-called replicated TRN-TST cross-validation that appears in the publication of Daetwyler et al. (2012), since some individuals can never be part of the training set. This algorithm, like the incomplete block cross-validation, is appropriate when we are interested in evaluating J lines in I blocks or environments and tries to mimic a prediction problem faced by breeders in incomplete field trials where lines are evaluated in some, but not all, target environments. The algorithm for constructing the TRN-TST sets is described by the following steps: Step 1. Calculate the total number of observations under study as $N = J \times I$; Step 2. Define the proportion of observations used for training and testing, that is, P_{TRN} and P_{TST}; Step 3. Calculate the size of the testing set $N_{TST} = N \times P_{TST}$; Step 4. Choose N_{TST} lines at

Table 4.2 TRN-TST data sets for $J = 10$ lines and $I = 3$ environments

Environment	Lines									
	L1	L2	L3	L4	L5	L6	L7	L8	L9	L10
1	TRN	TST	TRN	TRN	TST	TRN	TRN	TST	TRN	TRN
2	TST	TRN	TRN	TRN	TRN	TRN	TST	TRN	TST	TRN
3	TRN	TRN	TST	TST	TRN	TRN	TRN	TRN	TRN	TST

random without replacement if $J \geq N_{\text{TST}}$, and with replacement otherwise; Step 5. Each chosen line will then be assigned to one of the I environments chosen at random without replacement; Step 5. All the selected lines and environments will form the training set, while the lines and environments that were not chosen will form the corresponding testing set; and Step 6. Steps 1–5 are repeated depending on the number of TRN-TST partitions required (Lopez-Cruz et al. 2015). This CV is called CV2 in Jarquín et al. (2017). Next, we assume that we have ten lines or treatments and three environments or blocks that will form the corresponding training testing sets for only one partition: Step 1. The total number of observations under study is $N = 10 \times 3 = 30$; Step 2. We define $P_{\text{TRN}} = 0.7$ and $P_{\text{TST}} = 0.3$; Step 3. The size of the testing set is $N_{\text{TST}} = 30 \times 0.3 = 9$; Step 4. Since $J = 10 \geq N_{\text{TST}} = 9$, we selected the following lines at random without replacement: L1, L2, L3, L4, L5, L6, L7, L8, L9, and L10; and Step 5. Each chosen line was assigned to one of the $I = 3$ environments randomly chosen without replacement, as shown in Table 4.2. It is important to point out that this CV strategy only differs from the incomplete block cross-validation in the way the lines are allocated to blocks.

4.3.10 Other Options and General Comments on Cross-validation

It is important to highlight that when the data set is considerably large, it is better to randomly split it into three parts: a training set, a validation set (or tuning set), and a testing set. The training set and testing set are used as explained before, while the validation (tuning set) set is used to estimate the prediction error for model selection, which is the process of estimating the performance of different models in order to choose the best one, or to evaluate the chosen statistical machine learning model with a range of values of tuning hyperparameters to select the combination of hyperparameters with the best prediction performance and then use these hyperparameters (or best model) to evaluate the prediction performance in the testing set (Fig. 4.7). It is important to point out that Fig. 4.7 shows only one random split of the data in terms of the training, testing, and validation sets.

For example, assume that our data set has 50,000 rows (observations) and that we have decided to use 5000 of them for the testing set and another 5000 for the validation set. This means that 40,000 rows are left for the training set. At this point, we train our statistical machine learning model with each component of the

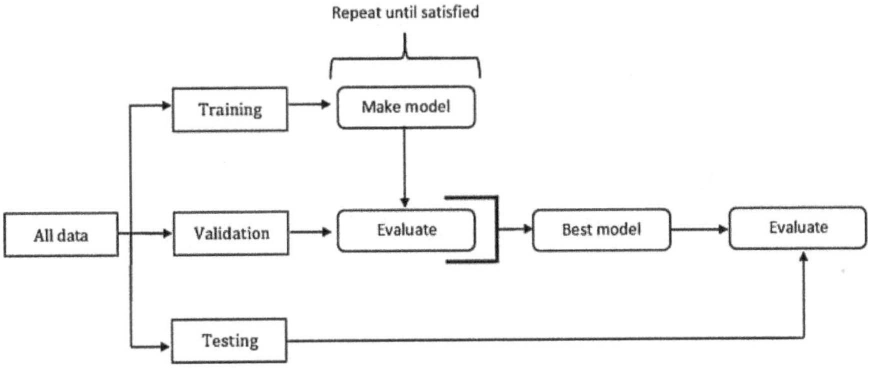

Fig. 4.7 Schematic representation of the training, validation (tuning), and testing sets proposed by Cook (2017)

grid of hyperparameters of the training set and evaluate the prediction performance on the validation set, as shown in the middle of Fig. 4.7. Finally, we will pick the best model (best subset of hyperparameters) in terms of prediction performance in the validation set and we are ready to evaluate the prediction performance in the testing set. Then, we will report the testing error on the testing set, as can be observed at the bottom of Fig. 4.7 (Cook 2017). Under this approach, the testing and validation sets have approximately the same size to guarantee a similar out-of-sample prediction performance. Since it is difficult to give general rules on how to choose the number of observations in each of the three parts, a typical number might be 50% for training, and 25% each for validation and testing (Hastie et al. 2008) or 70%, 15%, and 15% for training, validation, and testing, respectively. Another way to find the optimal setting of hyperparameters using the grid search, which is very common in deep learning, consists of picking values for each parameter from a finite set of options [e.g., number of epochs (100, 150, 200, 250, 300); batch sizes (25, 50, 75, 100, 125); number of layers (1, 2, 3, 4, 5); and types of activation functions (RELU, Sigmoid), . . .] and training the statistical machine learning model with every permutation of hyperparameter choices using the training set. Then, the combination of hyperparameters with the best prediction performance on the validation set is chosen, and we report the prediction performance of the best selected model (set of hyperparameters) in the testing set (Buduma 2017). The aforementioned examples use the validation data set as a proxy measure of the accuracy during the hyperparameter optimization process. However, it can also be used as a proxy measure of the accuracy for model selection, and instead of using a grid of hyperparameters, we can use a set of different statistical machine learning models; here, the best model is chosen instead of the best combination of hyperparameters. It is important to understand that when you have more than one random partition (training-testing-validation), as shown in Fig. 4.8, the same process provided in Fig. 4.7 is followed, however, the average of all the partitions is reported as a measure of prediction performance. Also, if more precision is required in the estimated prediction performance, you can repeat the process given in Fig. 4.8

<table>
<tr><td>Partition 1</td><td colspan="3">TRN</td><td>TST</td><td>VAL</td></tr>
<tr><td>Partition 2</td><td colspan="2">TRN</td><td>TST</td><td>VAL</td><td>TRN</td></tr>
<tr><td>Partition 3</td><td>TRN</td><td>TST</td><td>VAL</td><td colspan="2">TRN</td></tr>
<tr><td>Partition 4</td><td>TST</td><td>VAL</td><td colspan="3">TRN</td></tr>
<tr><td>Partition 5</td><td>VAL</td><td colspan="3">TRN</td><td>TST</td></tr>
</table>

Fig. 4.8 Five partitions of training-validation-testing

multiple times and report the average of all repetitions as a measure of prediction performance.

As mentioned above, this approach (Figs. 4.7 and 4.8) is used for a large data set however, in the case of a smaller data set, we suggest modifying this approach to avoid wasting too much training data in validation sets. This modification consists of performing an inner cross-validation approach since the training set is split into complementary subsets, and each model is trained against a different combination of these subsets, which is subsequently validated against the remaining parts. Once the model hyperparameters have been selected, a final model is trained with the whole training set (refitted) and the generalized prediction performance is measured on the testing set. This approach was applied by Montesinos-López et al. (2018a, b), who also split the original data into training and testing sets. Subsequently, each training set was split again and 80% of the data was used for training a grid of hyperparameters while the remaining 20% was used for validating (tuning) the prediction performance and selecting the best combination of hyperparameters with the best prediction performance. At this point, the deep learning algorithm was refitted with the whole training set, and with this they evaluated the out-of-sample prediction. They called the conventional training-testing partition outer cross-validation, while the split performed in each training set used for hyperparameter tuning was called inner cross-validation. It should be highlighted that there are no differences between outer and inner CV and training-validation-test. Finally, it should also be mentioned that any type of the cross-validation strategies mentioned in this section (random CV, *k*-fold CV, Bootstrap CV, IB CV, etc.,) can be used in both outer and inner CV, such as is the case with the five-fold CV.

4.4 Model Tuning

A hyperparameter is a parameter whose value is set before the learning process begins. Hyperparameters govern many aspects of the behavior of statistical machine learning models, such as their ability to learn features from data, the models' exhibited degree of generalizability in performance when presented with new data, as well as the time and memory cost of training the model, since different hyperparameters often result in models with significantly different performance. This means that tuning hyperparameter values is a critical aspect of the statistical machine learning training process and a key element for the quality of the resulting

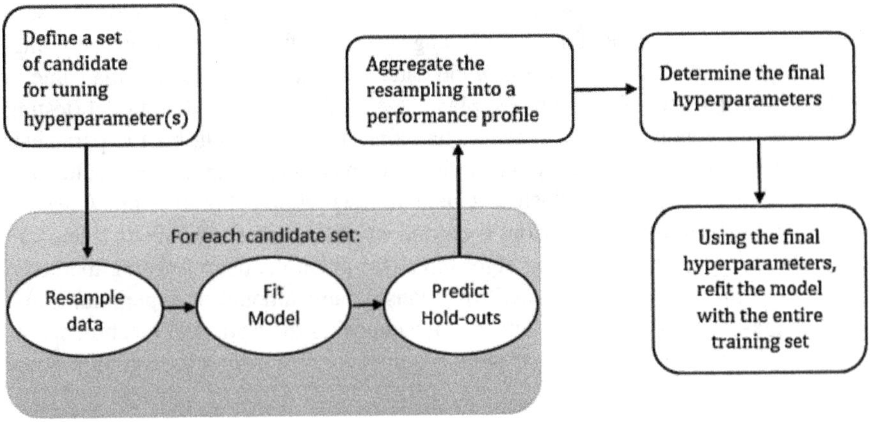

Fig. 4.9 Schematic representation of the tuning process proposed by Kuhn and Johnson (2013)

prediction accuracies. However, choosing appropriate hyperparameters is challenging (Montesinos-López et al. 2018a). Hyperparameter tuning finds the best version of a statistical machine learning model by running many training sets on the original data set using the algorithm and ranges of values of hyperparameters as specified. The hyperparameter values that provide the best performance in out-of-sample prediction evaluated by the chosen metric are then selected.

There are many ways of searching for the best hyperparameters. However, a general approach defines a set of candidate values for each hyperparameter. Each value of this set of candidate values is then applied with a resample of the training set of the chosen statistical machine learning method, where we aggregate all the hold-out predictions from which the best hyperparameters are chosen and refit the model with the entire set (Kuhn and Johnson 2013). A schematic representation of the tuning process proposed by Kuhn and Johnson (2013) is given in Fig. 4.9. It is important to highlight that this process should be performed correctly because when the same data are used for training and evaluating the prediction performance, the prediction performance obtained is extremely optimistic.

For example, suppose a breeder is interested in developing an algorithm to classify unseen plants as diseased or not diseased with an available training data set. The goal is to minimize the rate of misclassification or to maximize the percentage of cases correctly classified (PCCC). Also, assume that you are new to the world of statistical machine learning and that you only understand the k-nearest neighbor method. Since this algorithm depends only on the hyperparameter called the number of neighbors (k), the question is which value of k to choose in such a way that the prediction performance of this algorithm will be the best in the sample prediction of plants. To find the best value of the k hyperparameter, you must specify a range of values for k (for example, from 1 to 60 with increments of 1), then with a part of your training data set, called the training-inner (or tuning that corresponds to the training data in the inner loop) set, which is randomly selected. You proceed to evaluate the 60 values of k with the k-nearest neighbor method and evaluate the prediction performance in the remaining part of the training set (validation set). Next, you select the value of *k* from this range of values that best predicts (according, for example, to the PCCC) out-of-sample data (validation set) and use this value to perform the prediction of the unseen plants not used for training the model (testing set). This is a widely adopted practice that consists of searching for the parameter (usually through brute force loops) that yields the best performance over a validation set. However, the process illustrated here is very simple because the *k*-nearest neighbor model only depends on a unique hyperparameter; however, there are other statistical machine learning algorithms (for example, deep learning methods) where the tuning process is required for a considerable amount of hyperparameters. For this reason, we encourage caution when choosing the statistical machine learning algorithm, since the amount of work required for performing the tuning process depends on the chosen method.

4.4.1 Why Is Model Tuning Important?

Tuning the hyperparameters of the models is a key element to optimize your statistical machine learning model to perform well in out-of-sample predictions. The tuning process is more an art than a science because there is no unique formal scientific procedure available in the literature. Nowadays, the tuning process is trial and error that consists of implementing the statistical machine learning model many times with different values of the hyperparameters and then comparing its performance on the validation set in order to determine which set of hyperparameters results in the most accurate model; for the final implementation, the set of hyperparameters of the best model is used. As mentioned above, for the *k*-nearest neighbor classifier, we need to choose the number of neighbors (*k*) using the tuning process to obtain the optimal prediction performance of this algorithm, while for conventional Ridge regression, the parameter lambda (λ) is obtained by tuning to improve the out-of-sample predictions. These two statistical machine learning algorithms that we just mentioned need only one hyperparameter; however, other statistical machine learning methods may require more hyperparameters, as exemplified by deep learning models that require at least three hyperparameters (number of neurons, number of hidden layers, type of activation function, batch size, etc.). After tuning the required hyperparameters, the statistical machine learning model learns the parameters from the data to be used for the final prediction of the testing set. The choice of hyperparameters significantly influences the time required to train and test a statistical machine learning model. Hyperparameters can be continuous or of the integer type; for this reason, there are mixed-type hyperparameter optimization methods.

4.4.2 Methods for Hyperparameter Tuning (Grid Search, Random Search, etc.)

Manual tuning of statistical machine learning models is of course possible, but relies heavily on the user's expertise and understanding of the underlying problem.

Additionally, due to factors such as time-consuming model evaluations, nonlinear hyperparameter interactions in the case of large models that consist of tens or even hundreds of hyperparameters, manual tuning may not be feasible since it is equivalent to brute force. For this reason, the four most common approaches for hyperparameter tuning reported in the literature are (a) grid search, (b) random search, (c) Latin hypercube sampling, and (d) optimization (Koch et al. 2017).

In the *grid search* method, each hyperparameter of interest is discretized into a desired set of values to be studied where the models are trained and assessed for all combinations of the values across all hyperparameters (that is, a "grid"). Although fairly simple and straightforward to carry out, a grid search is appropriate when there are only a few values for a limited number of hyperparameters. However, although this is a comprehensive way of assessing different hyperparameter values, when there are many values for some or many hyperparameters, it quickly becomes quite costly due to the number of hyperparameters and the number of discrete levels of each. For example, in Ridge regression, this approach is implemented as follows: since λ is the hyperparameter to be tuned, we first propose, for example, a grid of 100 values for this hyperparameter from $\lambda = 10^{10}$ to $\lambda = 10^{-2}$; then we divide the training set into five inner training sets and five inner testing (tuning) sets, where each of the 100 values of the grid is fitted using the inner training sets and the testing error is evaluated with the inner testing sets. Then we get the average predicted test error and pick one value out of the 100 values of the grid that produces the best prediction performance. Next, we refit the statistical machine learning method to the whole training set using the picked value of λ, and finally perform the predictions for the testing set using the learned parameters of the training set with the best picked value of λ. In all the models with one hyperparameter, it is practical to implement the *grid search* method, but for example, in deep learning models, which many times require six hyperparameters to be tuned, if only three values are used for each hyperparameter, there are $3^6 = 729$ combinations that need to be evaluated, quickly becoming computationally impracticable.

A *random search* differs from a *grid search* in that rather than providing a discrete set of values to explore each hyperparameter, we determine a statistical distribution for each hyperparameter from which values may be randomly sampled. This affords a much greater chance of finding effective values for each hyperparameter. While Latin hypercube sampling is similar to the previous method, it is a more structured approach (McKay 1992) since it is an experimental design in which samples are exactly uniform across each hyperparameter but random in combinations. These so-called low-discrepancy point sets attempt to ensure that points are approximately equidistant from one another in order to fill the space efficiently. This sampling supports coverage across the entire range of each hyperparameter and is more likely to find good values of each hyperparameter.

The previous two methods for hyperparameter tuning are used to perform individual experiments by building models with various hyperparameter values and recording the model performance for each. Because each experiment is performed in isolation, this process is parallelized, but is unable to use the information from one experiment to improve the next experiment. Optimization methods, on the other hand, consist of sequential model-based optimization where the results of previous experiments are used to improve the sampling method of the next experiment. These methods are designed to make intelligent use of fewer evaluations and thus save on the overall computation time (Koch et al. 2017). Optimization algorithms that have been used in statistical machine learning generally for hyperparameter tuning include Broyden–Fletcher–Goldfarb–Shanno (BFGS) (Konen et al. 2011), covariance matrix adaptation evolution strategy (CMA-ES) (Konen et al. 2011), particle swarm (PS) (Renukadevi and Thangaraj 2014), tabu search (TS), genetic algorithms (GA) (Lorena and de Carvalho 2008), and more recently, surrogate-based Bayesian optimization (Dewancker et al. 2016). Also, recently the use of the response surface methodology has been explored for tuning hyperparameters in random forest models (Lujan-Moreno et al. 2018). However, the implementation of these optimization methods is not straightforward because it requires expensive computation; also, software development is required for implementing these algorithms automatically. There have been advances in this direction for some machine learning algorithms in the statistical analysis system (SAS), R and Python software (Koch et al. 2017). An additional challenge is the potential unpredictable computation expense of training and validating predictive models using different hyperparameter values. Finally, although it is challenging, the tuning process often leads to hyperparameter settings that are better than the default values, as it provides a heuristic validation of these settings, giving greater assurance that a model configuration with a higher accuracy has not been overlooked.

4.5 Metrics for the Evaluation of Prediction Performance

The quality of prediction performance of any statistical machine learning method in a given data set consists of evaluating how close the predicted values are to the true observed ones. In other words, the prediction performance quantifies the matching degree between the predicted response value for a given observation and the true response value for that observation (James et al. 2013). However, the metrics used for quantifying the prediction performance depend on the type of response variable under study; for this reason, we subsequently give the most popular metrics used for this goal for four types of response variables.

4.5.1 *Quantitative Measures of Prediction Performance*

Before implementing a statistical machine learning model, we assume that we have our training observation $\{(x_1, y_1), (x_2, y_2), \ldots, (x_n, y_n)\}$ and we estimate *f*, as $\widehat{f}$, with the chosen statistical machine learning model. Then we can make predictions for each of the response values (y_i) with $\widehat{f}(x_i)$ and compute the predicted values for each of the n observations in the training set; with these values we can calculate the ***mean square error*** (MSE) for the training data set as $E = \frac{1}{n}\left(\sum_{i=1}^{n}\left(y_i - \widehat{f}(x_i)\right)^2\right.$; however, what we really want to predict are the values for unseen test observations that were not used to train the statistical machine learning model. Assuming that the unseen testing set is equal to $\{(x_{n+1}, y_{n+1}), (x_{n+2}, y_{n+2}), \ldots, (x_{n+T}, y_{n+T})\}$, the MSE for the testing data set should be calculated as

$$\mathrm{MSE}_{\mathrm{TST}} = \frac{1}{T}\sum_{i=n+1}^{n+T}\left(y_i - \widehat{f}(x_i)\right)^2, \tag{4.1}$$

where $\widehat{f}(x_i)$ is the prediction that $\widehat{f}$ gives to the ith observation. The $\mathrm{MSE}_{\mathrm{TST}}$ with a lower value will have better predictions, which means that the predicted values are very close to the true observed values. Also, the square root of $\mathrm{MSE}_{\mathrm{TST}}$ can be used as a measure of prediction performance and is called root mean square error (RMSE).

Pearson's correlation coefficient is a very popular measure of prediction performance in plant breeding and can be calculated as

$$r_{\mathrm{TST}} = \frac{\sum_{i=n+1}^{n+T}\left(\widehat{f}(x_i) - \overline{\widehat{f}(x_i)}\right)(y_i - \overline{y}_i)}{\sqrt{\sum_{i=n+1}^{n+T}\left(\widehat{f}(x_i) - \overline{\widehat{f}(x_i)}\right)^2}\sqrt{\sum_{i=n+1}^{n+T}(y_i - \overline{y}_i)^2}}, \tag{4.2}$$

where $\overline{\widehat{f}(x_i)}$ is the average of the T predictions that conform to the testing set, and $\overline{y}_i$ is the average of the T true observed values. In this case, the closer that the predictions are to 1, the better the implemented statistical machine learning model will perform. It is important to point out that Pearson's correlation is defined between -1 and 1. However, to be convinced that the observed and predicted values match, it is a common practice to perform a scatter plot of predicted versus observed (or vice versa) values and when the observed and predicted values follow a straight line (45° diagonal line) from the bottom left corner to the top right corner, this indicates a perfect match between the observed and predicted values. For this reason, Pearson's correlation as a metric should be complemented with the intercept and slope, since the slope and intercept describe the consistency and the model bias, respectively (Smith and Rose 1995; Mesple et al. 1996). To obtain the slope and intercept, the observed values (as y) versus the predicted values (as x) are regressed and in addition to Pearson's correlation, these values should also be reported (slope and intercept). The expected intercept should be zero and the slope 1, if the correlation obtained between the observed and predicted values is high. It is important to avoid carrying out the regression in the opposite way, i.e., using predicted values as y's, and observed values as x's, since this leads to incorrect estimates of the slope and the y-intercept. This denotes that a spurious effect is added to the regression parameters when regressing predicted versus observed values and comparing them against the 1: 1 line. The user should also remember that underestimation of the slope and overestimation of the y-intercept increase as Pearson's correlation values decrease. We strongly recommend that scientists evaluate their models by regressing observed versus predicted values and test the significance of slope = 1 and intercept = 0 (Piñeiro et al. 2008). Finally, it is important to recall that the square of Pearson's correlation can also be used as a metric for measuring prediction performance since it represents the proportion of the total variance explained by the regression model and is called the coefficient of determination denoted as R^2.

Next, we present the mean absolute error (MAE) metric that measures the difference between two continuous variables (observed and predicted). The MAE can be calculated with the following expression:

$$\mathrm{MAE_{TST}} = \frac{1}{T}\sum_{i=n+1}^{n+T}\left|y_i - \widehat{f}(x_i)\right| \tag{4.3}$$

Below we present another metric used to evaluate the prediction performance of any statistical machine learning model; it was proposed by Kim and Kim (2016) and is called ***mean arctangent absolute percentage error*** (MAAPE) that is calculated as follows:

$$\mathrm{MAAPE_{TST}} = \frac{\sum_{i=n+1}^{n+T}\arctan\left(\left|\frac{y_i-\widehat{f}(x_i)}{y_i}\right|\right)}{T} \tag{4.4}$$

Although MAAPE is finite when the response variable (i.e., $y_i = 0$) equals zero, since it has a satisfactory trigonometric representation. However, because MAAPE's value is expressed in radians, it is less intuitive, in addition to being scale-free. This metric is a modification of the mean absolute percentage error (MAPE) which is problematic because it is undefined when the response variable is equal to zero ($y_i = 0$). MAAPE is also asymmetric since division by zero is defined and is not a problem. It is important to point out that there are other metrics for measuring prediction accuracy for continuous data, but we only presented the most popular ones.

The distinction between the training and test MSE is important since we are not interested in how well the statistical machine learning method performs in the training data set, due to the fact that our main goal is to perform accurate predictions in the unseen test data. For example, a plant breeder may be interested in developing an algorithm to predict disease resistance of a plant in new environments based on records that were collected in a set of environments. We can train the statistical machine learning method with the information collected in the set of environments, but the interest is not in how well the statistical machine learning method predicts in those previously collected environments. An environmental scientist can also be interested in predicting the average annual rainfall in a municipality in Mexico, using data from the last 20 years to train the model. Such measures could include sea surface temperature, time, the yearly rotation of the earth, among others. In this case, the scientist is really interested in predicting the average rainfall of the next 1 or 2 years, not in accurately predicting the years measured in the training set.

4.5.2 *Binary and Ordinal Measures of Prediction Performance*

The binary and ordinal response variables are very common in classification problems, where the goal is to predict which category something falls into. An example of a classification problem is analyzing financial data to determine if a client will be granted credit or not. Another example is analyzing breeding data to predict if an animal is at high risk for a certain disease or not. Below, we provide some popular metrics to evaluate the prediction performance of this type of data.

The first metric is called a ***confusion matrix***, which is a tool to visualize the performance of a statistical machine learning algorithm that is used in supervised learning for classifying categorical and binary data. Each column of the matrix represents the number of predictions in each class, while each row represents the instances in the real class. One of the benefits of confusion matrices is that they make it easy to determine whether the system is confusing classes. Table 4.3 shows a sample format of a confusion matrix of C classes.

With Eqs. (4.5–4.8) we can calculate the total number of false negatives (TFN), false positives (TFP), true negatives (TTN) for each class i, and the total true positives in the system, respectively:

$$\text{TFN}_i = \sum_{\substack{j=1 \\ j \neq i}}^{C} n_{ij} \tag{4.5}$$

Table 4.3 Confusion matrix with more than two classes

		Predicted values			
		Class 1	Class 2	...	Class C
Observed values	Class 1	n_{11}	n_{12}	...	n_{1C}
	Class 2	n_{21}	n_{22}	...	n_{2C}
	⋮	⋮	⋮	⋮	⋮
	Class C	n_{C1}	n_{C2}	...	n_{CC}

$$\mathrm{TFP}_i = \sum_{\substack{j=1 \\ j\neq i}}^{C} n_{ji} \tag{4.6}$$

$$\mathrm{TTN}_i = \sum_{\substack{j=1 \\ j\neq i}}^{C} \sum_{\substack{k=1 \\ k\neq i}}^{C} n_{ji} \tag{4.7}$$

$$\mathrm{TTP}_{\mathrm{all}} = \sum_{j=1}^{C} n_{jj} \tag{4.8}$$

Below, we define the sensitivity (S_e), precision (P), and specificity (S_p). The sensitivity indicates the ability of our statistical learning algorithm to determine the proportion of true positives that are correctly identified by the test. The precision is the proportion of correct classification of our statistical machine learning model and represents the proportion of cases correctly classified, while the specificity is the ability of our statistical machine learning model to classify the true negative cases, that is, the specificity is the proportion of true negatives that are correctly identified by the test. Under the "one-versus-all basis," where each category is compared with the composed information of the remaining categories, we provide the expressions for computing the generalized precision, sensitivity, and specificity for each class i:

$$P_i = \frac{\mathrm{TTP}_{\mathrm{all}}}{\mathrm{TTP}_{\mathrm{all}} + \mathrm{TFP}_i} \tag{4.9}$$

$$S_{ei} = \frac{\mathrm{TTP}_{\mathrm{all}}}{\mathrm{TTP}_{\mathrm{all}} + \mathrm{TFN}_i} \tag{4.10}$$

$$S_{pi} = \frac{\mathrm{TTN}_{\mathrm{all}}}{\mathrm{TTN}_{\mathrm{all}} + \mathrm{TFP}_i} \tag{4.11}$$

$$\mathrm{pCCC} = \frac{\mathrm{TTN}_{\mathrm{all}}}{\sum_{i=1}^{C} \sum_{j=1}^{C} n_{ij}} \tag{4.12}$$

The term pCCC denotes the ***proportion of cases correctly classified***, which is a measure of the overall accuracy, and when multiplied by 100, denotes the percentage of cases correctly classified. Many times, this is the only metric reported for measuring prediction performance in multi-class problems. However, PCCC alone is sometimes quite misleading as there may be a model with relatively "high" accuracy, but it predicts the "unimportant" class labels fairly accurately (e.g., "unknown bucket"). However, the model may be making all sorts of mistakes on the classes that are actually critical to the application. This problem is serious when in the input data the number of samples of different classes is very unbalanced. For

Table 4.4 Confusion matrix with two classes

		Predicted values		
		True	False	Sum
Observed values	True	tp	fn	tp + fn
	False	fp	tn	fp + tn
	Sum	tp + fp	fn + tn	n

tp denotes true positives, fp denotes false positives, fn denotes false negatives, tn denotes true negatives, and n denotes total number of individuals

example, if there are 990 samples of class 1 and only 10 of class 2, the classifier can easily have a bias toward class 1. If the classifier classifies all samples as class 1, its accuracy will be 99%. This does not mean that it is an appropriate classifier, as it had a 100% error when classifying the samples of class 2. For this reason, reporting this metric with those reported in Eqs. (4.9–4.11) is recommended in order to have a better picture of the prediction performance of any statistical machine learning method (Ratner 2017). Also, it is important to highlight that when the problem only has two classes, the confusion matrix is reduced to Table 4.4.

From Table 4.4 the PCCC is calculated as $\frac{\text{tp+tn}}{n}$, while the $S_e = \frac{\text{tp}}{\text{tp+fn}}$, $S_p = \frac{\text{tn}}{\text{tn+fp}}$, and $P = \frac{\text{tp}}{\text{tp+fp}}$. Also, when there are only two classes, González-Camacho et al. (2018) suggest calculating the ***Kappa coefficient*** (κ) or Cohen's Kappa, which is defined as

$$\kappa = \frac{P_0 - P_e}{1 - P_e},$$

where P_0 is the agreement between observed and predicted values and is computed by the PCCC described above for two classes; P_e is the probability of agreement calculated as $P_e = \frac{\text{tp+fn}}{n} \times \frac{\text{tp+fp}}{n} + \frac{\text{fp+tn}}{n} \times \frac{\text{fn+tn}}{n}$, where fp is the number of false positives, and fn is the number of false negatives (Table 4.4). This statistic can take on values between -1 and 1; a value of 0 means there is no agreement between the observed and predicted classes, while a value of 1 indicates perfect agreement between the model prediction and the observed classes. Negative values indicate that a prediction may be incorrect; however large negative values seldom occur when working with predictive models. Depending on the context, a Kappa value from 0.30 to 0.50 indicates reasonable agreement (Kuhn and Johnson 2013). The Kappa coefficient is appropriate when data are unbalanced, because it estimates the proportion of cases that were correctly identified by taking into account coincidences expected from chance alone (Fielding and Bell 1997). It is important to point out that this statistic was originally designed to assess the agreement between two raters (Cohen 1960).

Another popular metric for binary data is the ***Area Under the receiver operating characteristic*** Curve (AUC–ROC) and it ranks the positive predictions higher than the negative. The ROC curve is defined as a plot of 1 − specificity or false positive rate (FPR) as the x-axis versus its model sensitivity as the y-axis. For a given set of

Fig. 4.10 ROC curve for the synthetic data

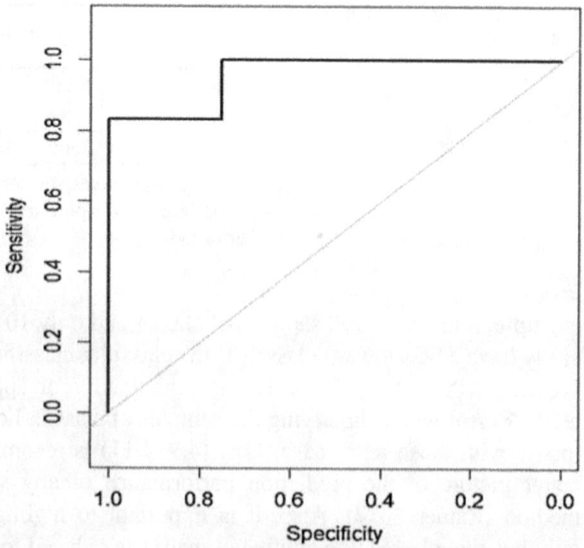

thresholds τ, it is an effective method for evaluating the quality or performance of diagnostic tests, and is widely used in statistical machine learning to evaluate the prediction performance of learning algorithms. Since the mathematical construction of this metric is not required in this book, we illustrate its calculation with one simple example. Assume that the observed (y), predicted probabilities (*pi*) and predicted values ($\hat{y}$) obtained after implementing a statistical machine learning model are $y = \{1, 0, 1, 1, 0, 0, 1, 1, 0, 1\}$, $pi = \{0.6, 0.55, 0.8, 0.78, 0.3, 0.42, 0.9, 0.45, 0.3, 0.88\}$, and $\hat{y} = \{1, 1, 1, 1, 0, 0, 1, 0, 0, 1\}$; then, using the following R code, we can obtain many metrics for binary data (Fig. 4.10):

```
#######################Libraries required#######################
  library(caret)
  library(pROC)
  ##Observed (y), predicted probability (pi) and predicted values of
synthetic data##
  y=c(1, 0,1, 1, 0, 0, 1, 1, 0, 1)
  pi=c(0.6, 0.55, 0.8, 0.78, 0.3, 0.42, 0.9, 0.45, 0.3, 0.88)
  yhat=c(1, 1, 1, 1, 0, 0, 1, 0, 0, 1)
  xtab <- table(y,yhat)
  confusionMatrix(xtab)

  plot.roc(y,pi) #####This make the ROC curve plot

  Confusion Matrix and Statistics

    yhat
  y  0 1
   0 3 1
   1 1 5
```

```
                Accuracy : 0.8
                  95% CI : (0.4439, 0.9748)
     No Information Rate : 0.6
     P-Value [Acc > NIR] : 0.1673

                   Kappa : 0.5833
  Mcnemar's Test P-Value : 1.0000

             Sensitivity : 0.7500
             Specificity : 0.8333
          Pos Pred Value : 0.7500
          Neg Pred Value : 0.8333
              Prevalence : 0.4000
          Detection Rate : 0.3000
    Detection Prevalence : 0.4000
       Balanced Accuracy : 0.7917

        'Positive' Class : 0
```

It is important to point out that when there are more than two classes, these metrics (accuracy, sensitivity, specificity, etc.) can be calculated on a "one-versus-all" basis that consists of using each class versus the pool of the remaining classes, as was illustrated for the confusion matrix with more than two classes (James et al. 2013).

When the statistical machine learning model discriminates correctly between the two groups, it produces a curve that coincides with the left and top sides of the plot. Under this scenario, the perfect model would have 100% sensitivity and specificity, and the ROC curve would be a single step between (0,0) and (0,1) and would remain constant from (0,1) to (1,1); this implies that the area under the ROC curve of the model would be 1. In general, the larger the area under the ROC curve, the better the model in terms of prediction performance. On the other hand, a completely useless statistical machine learning algorithm would give a straight line (45° diagonal line) from the bottom left corner to the top right corner of the plot. Different statistical machine learning models or the same model with different training sets or hyperparameters can be compared by superimposing their ROC curves in the same graph. In practice, most of the time the values in the two groups overlap, so the curve often lies between these extremes.

From this ROC curve, we can obtain a global assessment of the prediction performance of the statistical machine learning method by measuring the area under the receiver operating characteristic curve. This area is equal to the probability that a random individual (person) of the sample with the presence of the target has a higher value of the measurement than a random individual without the target.

When a statistical machine learning model is unable to discriminate between the positive and negative classes, this means that it has low discriminatory power. Therefore, only in statistical machine learning models that had good discriminatory power we can be confident of the predictions they provide and furthermore those models that provide a curve that lies considerably above the curve will be better.

Matthews correlation coefficient (MCC). Introduced in 1975 by Brian Matthews (1975) and regarded by many scientists as the most informative score that connects all four measures in a confusion matrix, the Matthews Correlation Coefficient is typically used in statistical machine learning to measure the quality of binary classifications and it is particularly useful when there is a significant imbalance in class sizes (data). MCC is calculated according to the following expression:

$$\text{MCC} = \frac{\text{tp} \times \text{tn} - \text{fp} \times \text{fn}}{\sqrt{(\text{tp}+\text{fp}) \times (\text{tp}+\text{fn}) \times (\text{tn}+\text{fp}) \times (\text{tn}+\text{fn})}} \tag{4.13}$$

If any of the denominator terms equals zero in (4.13) it will be set to 1 and MCC becomes zero, which has been shown to be the correct limiting value. It returns a value between −1 and 1, where 1 means a perfect prediction, 0 means no better than random and −1 means a total disagreement between predicted and observed values.

Next, we present the ***Brier score*** (Brier 1950) for categorical or binary data that can be computed as

$$\text{BS} = T^{-1} \sum_{i=n+1}^{n+T} \sum_{c=1}^{C} (\widehat{\pi}_{ic} - d_{ic})^2, \tag{4.14}$$

where $\widehat{\pi}_i$ denotes the estimated probabilities (predictive distribution) derived from the estimated model for observation i and d_{ic} takes a value of 1 if the categorical response observed for individual i falls into category c; otherwise, $d_{ic} = 0$. The range of BS in Eq. (4.14) for categorical data is between 0 and 2. For this reason, we suggest dividing by 2, that is, BS/2, to obtain the Brier score bound between 0 and 1; lower scores imply better predictions (Montesinos-López et al. 2015a, b).

Finally, we describe the use of negative log-likelihood (MLL) to evaluate the prediction performance. This metric has the characteristic that better forecasts have lower values and for this reason, it is analogous to the MSE. For categorical data,

$$\text{MLL} = -\frac{1}{T}\left[\sum_{i=n+1}^{n+T} \sum_{c=1}^{C} 1\{y_i = k\} \log(\widehat{\pi}_{ic})\right],$$

where $1\{y^{(i)} = k\}$ is an indicator variable taken the value of 1 when the ith observation is assigned to category c, for $c = 1, 2, \ldots, C$ takes place in the ith observation. When the data are binary, the MLL is reduced to $\text{MLL} = -\frac{1}{T}\left[\sum_{i=n+1}^{n+T} [y_i \log(\widehat{\pi}_i) + (1-y_i)\log(1-\widehat{\pi}_i)]\right]$. Following, we provide some advantages of using the MML as a measure of prediction performance: (a) it has a simple definition that, from a purely intuitive point of view, seems to be a reasonable basis on which to compare forecasts; (b) it is mathematically optimal in the sense that estimates of parameters of calibration models fitted by maximizing the likelihood are usually the most accurate possible estimates (see Cassella and Berger 2002); (c) it is a generalization to probabilistic forecasts of the most commonly used skill score for single forecasts; (d) the properties of the likelihood have been studied at great length over the last 90 years, and as such is well understood; (e) it is both a measure of resolution and reliability; (f) likelihood can be used for both calibration and assessment: this creates consistency between these two operations; (g) use of the likelihood also creates consistency with other statistical modeling activities, since most other statistical modeling uses the likelihood, which is important in cases where the use of forecasts is simply a small part of a larger statistical modeling effort, as is the case of our particular business; (h) likelihood can be used for all types of response variables; and (i) likelihood can be used to compare multiple leads, multiple variables, and multiple locations at the same time in a sensible way with a single score even when these leads, variables, and locations are cross-correlated.

4.5.3 Count Measures of Prediction Performance

Spearman's correlation and the MML are recommended to measure the prediction performance for count data.

For the application of the ***Spearman's correlation***, the formula given in Eq. (4.2) for Pearson's correlation can be used; however, instead of using the observed and predicted values directly, these are replaced by their corresponding ranks. For example, assuming that the observed and predicted values are $y = \{15, 9, 12, 27, 6, 3, 36, 15, 21, 30\}$ and $\widehat{y} = \{20, 17, 24, 25, 3, 3, 34, 22, 21, 33\}$, we can thus show how to get the rank for the observed values: $\text{rango}_y = \{5, 3, 4, 8, 2, 1, 10, 6, 7, 9\}$. However, in this vector, observations 1 and 8 are the same, and as such, their positions are added and divided by two, that is, $\frac{5+6}{2} = 5.5$. Therefore, the final rank for the observed values is $\text{rango}_y = \{5.5, 3, 4, 8, 2, 1, 10, 5.5, 7, 9\}$. Now the range for the predicted values is $\text{rango}_{\widehat{y}} = \{4, 3, 7, 8, 1, 2, 10, 6, 5, 9\}$, but again, since values 5 and 6 are the same, we add their ranges, and as this is repeated twice, it is divided by two and we get $\frac{1+2}{2} = 1.5$. Therefore, the final range of the predicted values is $\text{rango}_{\widehat{y}} = \{4, 3, 7, 8, 1.5, 1.5, 10, 6, 5, 9\}$. Finally, to obtain the Spearman correlation, we used the expression given in Eq. (4.2) for Pearson's correlation, and instead of using the original observed and predicted values, we used rango_y and $\text{rango}_{\widehat{y}}$. The interpretation of this metric is equal to that of the Pearson correlation, that is, when it is closer to 1, the prediction performance of the implemented statistical learning method is better. It should be noted that when the number of repeated values in the observed and predicted values is greater than two, the adjusted range is the sum of the repeated ranges divided by the number of repeated values; this new range is then given to the repeated values. In this case, it is also important to regress the observed versus the predicted values to obtain the intercept and slope using the ranges of the observed and predicted values.

It is also possible to use the MLL criteria to assess the prediction performance for count data; however, the new expression is now based on minus the log-likelihood of a Poisson distribution, which is equal to

$$\mathrm{MLL} = \frac{1}{T} \sum_{i=n+1}^{n+T} \left[-\widehat{f}(x_i) + y_i \log \left(\widehat{f}(x_i) \right) \right]$$

Again, when the values of MLL are lower, the observed and predicted values are closer to one another.

References

Brier GW (1950) Verification of forecasts expressed in terms of probability. Mon Weather Rev 78: 1–3

Buduma M (2017) Fundamentals of deep learning, 1st edn. O'Reilly, Sabastopol, CA

Burger SV (2018) Introduction to machine learning with R. Rigorous mathematical analysis, 1st edn. O'Reilly, Sabastopol, CA

Cassella G, Berger RL (2002) Statistical inference. Duxbury, Belmont, CA

Cohen J (1960) A coefficient of agreement for national data. Educ Psychol Meas 20:37–46

Cook D (2017) Practical machine learning with H_2O. O'Reilly Media, Inc, Sabastopol, CA

Daetwyler HD, Calus MPL, Pong-Wong R, de los Campos G, Hickey JM (2012) Genomic prediction in animals and plants: simulation of data, validation, reporting and benchmarking. Genetics 193:347–365

Dewancker I, McCourt M, Clark S, Hayes P, Johnson A, Ke G (2016) A stratified analysis of Bayesian optimization methods. arXiv:1603.09441v1

Efron B (1983) Estimating the error rate of a prediction rule: improvement on cross-validation. J Am Stat Assoc 78(382):316–331

Fielding AH, Bell JF (1997) A review of methods for the assessment of prediction errors in conservation presence/absence models. Environ Conserv 24:38–49. https://doi.org/10.1017/S0376892997000088

González-Camacho JM, Ornella L, Pérez-Rodríguez P, Gianola D, Dreisigacker S, Crossa J (2018) Applications of machine learning methods to genomic selection in breeding wheat for rust resistance. Plant Genome 11(2):1–15. https://doi.org/10.3835/plantgenome2017.11.0104

Hagerty MR, Srinivasan S (1991) Comparing the predictive powers of alternative multiple regression models. Psychometrika 56:77–85. MR1115296

Hastie T, Tibshirani R, Friedman J (2008) The elements of statistical learning: data mining, inference, and prediction, Springer series in statistics, 2nd edn. Springer, New York

James G, Witten D, Hastie T, Tibshirani R (2013) An introduction to statistical learning: with applications in R. Springer, New York

Jarquín D, Lemes da Silva C, Gaynor RC, Poland J, Fritz AR et al (2017) Increasing genomic-enabled prediction accuracy by modeling genotype × environment interactions in Kansas wheat. Plant Genome 10(2):1–15. https://doi.org/10.3835/plantgenome2016.12.0130

Kim S, Kim H (2016) A new metric of absolute percentage error for intermittent demand forecasts. Int J Forecast 32(3):669–679

Koch P, Wujek B, Golovidov O, Gardner S (2017) Automated hyperparameter tuning for effective machine learning. In: Proceedings of the SAS global forum 2017 conference. SAS Institute Inc, Cary, NC. http://support.sas.com/resources/papers/proceedings17/SAS514-2017.pdf

Konen W, Koch P, Flasch O, Bartz-Beielstein T, Friese M, Naujoks B (2011) Tuned data mining: a benchmark study on different tuners. In: Proceedings of the 13th annual conference on genetic and evolutionary computation (GECCO-2011). SIGEVO/ACM, New York

Kuhn M, Johnson K (2013) Applied predictive modeling. Springer, New York

Lopez-Cruz M, Crossa J, Bonnett D, Dreisigacker S, Poland J, Jannink J-L, Singh RP, Autrique E, de los Campos, G. (2015) Increased prediction accuracy in wheat breeding trials using a marker × environment interaction genomic selection method. G3 5(4):569–582

Lorena AC, de Carvalho ACPLF (2008) Evolutionary tuning of SVM parameter values in multiclass problems. Neurocomputing 71:3326–3334

Lujan-Moreno GA, Howard PR, Rojas OG, Montgomery DC (2018) Design of experiments and response surface methodology to tune machine learning hyperparameters, with a random forest case-study. Expert Syst Appl 109:195–205

Matthews BW (1975) Comparison of the predicted and observed secondary structure of T4 phage lysozyme. Biochim Biophys Acta Protein Struct 405:442–451

McKay MD (1992) Latin hypercube sampling as a tool in uncertainty analysis of computer models. In: Swain JJ, Goldsman D, Crain RC, Wilson JR (eds) Proceedings of the 24th conference on winter simulation (WSC 1992). ACM, New York, pp 557–564

Mesple F, Troussellier M, Casellas C, Legendre P (1996) Evaluation of simple statistical criteria to qualify a simulation. Ecol Model 88:9–18

Montesinos-López OA, Montesinos-López A, Pérez-Rodríguez P, de los Campos G, Eskridge KM, Crossa J (2015a) Threshold models for genome-enabled prediction of ordinal categorical traits in plant breeding. G3 5(1):291–300

Montesinos-López OA, Montesinos-López A, Crossa J, Burgueño J, Eskridge K (2015b) Genomic-enabled prediction of ordinal data with Bayesian logistic ordinal regression. G3 5(10):2113–2126. https://doi.org/10.1534/g3.115.021154

Montesinos-López A, Montesinos-López OA, Gianola D, Crossa J, Hernández-Suárez CM (2018a) Multi-environment genomic prediction of plant traits using deep learners with a dense architecture. G3 8(12):3813–3828. https://doi.org/10.1534/g3.118.200740

Montesinos-López OA, Montesinos-López A, Crossa J, Gianola D, Hernández-Suárez CM et al (2018b) Multi-trait, multi-environment deep learning modeling for genomic-enabled prediction of plant traits. G3 8(12):3829–3840. https://doi.org/10.1534/g3.118.200728

Piñeiro G, Perelman S, Guerschman JP, Paruelo JM (2008) How to evaluate models: observed vs. predicted or predicted vs. observed? Ecol Model 216:316–322

Ratner B (2017) Statistical and machine-learning data mining. Techniques for better predictive modelling and analysis of big data, 3rd edn. CRC Press Taylor & Francis Group, Boca Raton, FL

Renukadevi NT, Thangaraj P (2014) Performance analysis of optimization techniques for medical image retrieval. J Theor Appl Inf Technol 59:390–399

Shalev-Shwartz S, Ben-David S (2014) Understanding machine learning from theory to algorithms. Cambridge University press, New York

Shmueli G (2010) To explain or to predict? Stat Sci 25(3):289–310

Smith EP, Rose KA (1995) Model goodness-of-fit analysis using regression and related techniques. Ecol Model 77:49–64

Wu S, Harris T, Mcauley K (2007) The use of simplified or misspecified models: linear case. Canad J Chem Eng 85:386–398

5

LMM: Estimation and Genomic Prediction

5.1 General of Linear Mixed Models

Linear mixed models (LMM) are flexible extensions of linear models in which fixed and random effects enter linearly into the model. This is useful in many disciplines to model repeated, longitudinal, or clustered observations, in which random effects are introduced to help capture correlation or/and random variation among observations in the same group of individuals. Random effects are random values associated with levels of a random factor, and often represent random deviations from the population mean and linear relationships described by fixed effects (Pinheiro and Bates 2000; West et al. 2014).

The first formulation of a linear mixed model was applied in the field of astronomy to analyze repeated telescopic observations made at various hourly intervals over a range of nights (West et al. 2014). The mixed model approach is often called by various names, depending on the discipline in which it is applied. For example, in the social sciences, this approach is known as a multilevel or hierarchical model that is often used to flexibly measure the different levels of grouping present in the data structure (e.g., an impact evaluation of a new teaching method, survey of job satisfaction, education applications, etc.) (Goldstein 2011; Speelman et al. 2018; Finch et al. 2019).

Other application areas can be found in medicine (health care research, Leyland and Goldstein 2001; Brown and Prescott 2014), agriculture, ecology, industry, and animal science, where this model is often referred to as the random effects or mixed-effects model (Pinheiro and Bates 2000; Raudenbush and Bryk 2002; Meeker et al. 2011; Zuur et al. 2009). Specifically, now there is an increasing number of applications of this model in genomic selection for plant and animal breeding, where molecular markers obtained by genotyping-by-sequencing or other technologies are used to predict breeding values for non-phenotyped lines to select candidate lines prior to phenotypic evaluation (Meuwissen et al. 2001; Poland et al. 2012; Cabrera-Bosquet et al. 2012; Araus and Cairns 2014; Crossa et al. 2017;

Covarrubias-Pazaran et al. 2018; Wang et al. 2018; Cappa et al. 2019). However, the use of this model in animal science can be traced back to Henderson (1950).

The general univariate linear mixed model (Harville 1977) is provided by the formula

$$Y = X\beta + Zb + \epsilon, \tag{5.1}$$

where Y is the $n \times 1$ random response vector, X is the $n \times (p + 1)$ design matrix for the fixed effects, $\beta = (\beta_0, \beta_1, \ldots, \beta_p)^T$ is the $(p + 1) \times 1$ coefficient vector of fixed effects, b is a $q \times 1$ vector of random effects, Z is the associate matrix design for the random effects, and ϵ is the $n \times 1$ vector of random errors. It assumes that ϵ is a random vector with a mean vector of $\mathbf{0}$ and a variance–covariance matrix R, b is a random vector with a mean of $\mathbf{0}$ and variance–covariance matrix D, and a null variance–covariance matrix between ϵ and b, $\text{Cov}(\epsilon, b) = \mathbf{0}_{n \times q}$. In genomic applications, b often includes the genotypic effects and genotype $\times$ environment interaction effects, while X may contain information about environment covariates and other related information.

Note that under this model, $E(Y) = X\beta$ and the variance–covariance matrix of the response vector is $\text{Var}(Y) = ZDZ^T + R$.

5.2 Estimation of the Linear Mixed Model

5.2.1 Maximum Likelihood Estimation

One method typically used for the estimation of the parameters of the LMM is the maximum likelihood approach. For the estimation under an LMM, the random errors and the random effects components are needed. Assuming that $\epsilon \sim N_n(\mathbf{0}, R)$, and $b \sim N_q(\mathbf{0}, D)$, with R and D positive semi-defined matrices, the marginal distribution of the response vector Y is $N_n(X\beta, ZDZ^T + R)$, and so the likelihood of the parameters is given by

$$L(\beta, D, R; y) = \frac{|V|^{-\frac{1}{2}}}{(2\pi)^{\frac{n}{2}}} \exp\left[-\frac{1}{2}(y - X\beta)^T V^{-1}(y - X\beta)\right], \tag{5.2}$$

where $V = Z^T DZ + R$ is the marginal variance of Y.

The maximum likelihood estimators (MLE) of the parameters, β, D, and R, are the values that maximize the likelihood function (5.2) (Searle et al. 2006; Stroup 2012), but due to the fact that no explicit formulas to estimate these parameters exist, numerical methods such as Newton–Raphson and Fisher Scoring are used. See details for implementing these methods in Jennrich and Sampson (1976) for the case where: D is a block diagonal with a submatrix in each diagonal of the form $\sigma_j^2 A_j$, where A_j is a known matrix and σ_j^2 is the variance component parameter to be estimated in this case; while for $\boldsymbol{R} = \sigma^2\boldsymbol{C}$, where $\boldsymbol{C}$ is a known matrix, only σ^2 should be estimated. For a more general explanation, see Jennrich and Schluchter (1986), and for an improvement of the algorithms proposed by these authors, consult Lindstrom and Bates (1988).

Another numerical method that can be used to obtain the MLE is the expected maximization (EM) algorithm, which is conceptually a simple algorithm for parameter estimation in this model (Laird and Ware 1982). This algorithm is an iterative numerical method to obtain maximum likelihood in the context of missing or hidden data (Borman 2004). The EM algorithm is described for the case where $\boldsymbol{R} = \sigma^2\boldsymbol{I}_n$, and where $\boldsymbol{I}_n$ is the identity matrix of dimension n. This algorithm, for some specific variance–covariance matrices of random effects, as described and used below, can be implemented using the sommer R package (Covarrubias-Pazaran 2016, 2018), which provides two additional algorithms available to obtain the MLE of the parameters in this same model.

5.2.1.1 EM Algorithm

The likelihood for complete data, $\boldsymbol{y}$ and $\boldsymbol{b}$, is given by

$$\begin{aligned} f_{\boldsymbol{Y},\boldsymbol{b}}(\boldsymbol{y},\boldsymbol{b}) &= f_{\boldsymbol{Y}|\boldsymbol{b}}(\boldsymbol{y}|\boldsymbol{b})f_{\boldsymbol{b}}(\boldsymbol{b}) \\ &= \frac{|\boldsymbol{D}|^{-\frac{1}{2}}}{(2\pi\sigma^2)^{\frac{n}{2}}}\exp\left[-\frac{1}{2\sigma^2}(\boldsymbol{y}-\boldsymbol{X\beta}-\boldsymbol{Zb})^{\mathrm{T}}(\boldsymbol{y}-\boldsymbol{X\beta}-\boldsymbol{Zb})-\frac{1}{2}\boldsymbol{b}^{\mathrm{T}}\boldsymbol{D}^{-1}\boldsymbol{b}\right] \end{aligned}$$

As such, the log-likelihood for the complete data, $\boldsymbol{y}$ and $\boldsymbol{b}$, is given by

$$\begin{aligned} \ell_c(\boldsymbol{\beta},\boldsymbol{\theta};\boldsymbol{y},\boldsymbol{b}) &= \log\left[f_{\boldsymbol{Y},\boldsymbol{b}}(\boldsymbol{y},\boldsymbol{b})\right] \\ &= -\frac{n}{2}\log\left(2\pi\sigma^2\right)-\frac{1}{2\sigma^2}(\boldsymbol{y}-\boldsymbol{X\beta}-\boldsymbol{Zb})^{\mathrm{T}}(\boldsymbol{y}-\boldsymbol{X\beta}-\boldsymbol{Zb}) \\ &\quad -\frac{1}{2}\boldsymbol{b}^{\mathrm{T}}\boldsymbol{D}^{-1}\boldsymbol{b}-\frac{1}{2}\log\left(|\boldsymbol{D}|\right), \end{aligned}$$

where $\boldsymbol{\theta}$ is the vector parameter that defines the variance–covariance matrix of the random effects ($\boldsymbol{D}$) and the random vector ($\boldsymbol{R}$). Some specific examples are given below.

E Step

Because $E(\boldsymbol{u}^{\mathrm{T}}\boldsymbol{A}\boldsymbol{u}) = \mathrm{tr}[\boldsymbol{A}\mathrm{Var}(\boldsymbol{u})] + E(\boldsymbol{u})^{\mathrm{T}}\boldsymbol{A}E(\boldsymbol{u})$ and $\boldsymbol{b} \mid \boldsymbol{Y}=\boldsymbol{y} \sim N_q\left(\widetilde{\boldsymbol{b}},\widetilde{\boldsymbol{D}}\right)$ (see Appendix 1), given the current values of the parameters $\boldsymbol{\beta}_{(t)}$ and $\boldsymbol{\theta}_{(t)}$, the conditional expected value of the complete likelihood, $\ell_c(\boldsymbol{\beta},\boldsymbol{\theta};\boldsymbol{y},\boldsymbol{b})$, is given by [Step (E)]:

$$\begin{aligned}
Q\left(\boldsymbol{\beta},\boldsymbol{\theta}|\boldsymbol{\beta}_{(t)},\boldsymbol{\theta}_{(t)},\right) &= E_{\boldsymbol{b}|\boldsymbol{y}}[\ell_c(\boldsymbol{\beta},\boldsymbol{b},\boldsymbol{\theta};\boldsymbol{y})] \\
&= -\frac{n}{2}\log\left(2\pi\sigma^2\right) - \frac{1}{2\sigma^2}\operatorname{tr}\left(\boldsymbol{Z}\widetilde{\boldsymbol{D}}_{(t)}\boldsymbol{Z}^{\mathrm{T}}\right) - \frac{1}{2\sigma^2}\left(\boldsymbol{y}-\boldsymbol{X\beta}-\boldsymbol{Z}\widetilde{\boldsymbol{b}}_{(t)}\right)^{\mathrm{T}}\left(\boldsymbol{y}-\boldsymbol{X\beta}-\boldsymbol{Z}\widetilde{\boldsymbol{b}}_{(t)}\right) \\
&\quad -\frac{1}{2}\operatorname{tr}\left(\boldsymbol{D}^{-1}\widetilde{\boldsymbol{D}}_{(t)}\right) - \frac{1}{2}\widetilde{\boldsymbol{b}}_{(t)}^{\mathrm{T}}\boldsymbol{D}^{-1}\widetilde{\boldsymbol{b}}_{(t)} - \frac{1}{2}\log\left(|\boldsymbol{D}|\right) \\
&= -\frac{n}{2}\log\left(2\pi\sigma^2\right) - \frac{1}{2\sigma^2}\left[\operatorname{tr}\left(\boldsymbol{Z}\widetilde{\boldsymbol{D}}_{(t)}\boldsymbol{Z}^{\mathrm{T}}\right) + \left(\boldsymbol{y}-\boldsymbol{X\beta}-\boldsymbol{Z}\widetilde{\boldsymbol{b}}_{(t)}\right)^{\mathrm{T}}\left(\boldsymbol{y}-\boldsymbol{X\beta}-\boldsymbol{Z}\widetilde{\boldsymbol{b}}_{(t)}\right)\right] \\
&\quad -\frac{1}{2}\left\{\operatorname{tr}\left[\left(\widetilde{\boldsymbol{D}}_{(t)}+\widetilde{\boldsymbol{b}}_{(t)}\widetilde{\boldsymbol{b}}_{(t)}^{\mathrm{T}}\right)\boldsymbol{D}^{-1}\right] + \log\left(|\boldsymbol{D}|\right)\right\}
\end{aligned}$$

where $\widetilde{\boldsymbol{D}}_{(t)} = \left(\boldsymbol{D}_{(t)}^{-1} + \sigma_{(t)}^{-2}\boldsymbol{Z}^{\mathrm{T}}\boldsymbol{Z}\right)^{-1}$, $\widetilde{\boldsymbol{b}}_{(t)} = \sigma_{(t)}^{-2}\widetilde{\boldsymbol{D}}_{(t)}\boldsymbol{Z}^{\mathrm{T}}\left(\boldsymbol{y}-\boldsymbol{X}\boldsymbol{\beta}_{(t)}\right)$, and $\boldsymbol{\beta}_{(t)}$, $\sigma_{(t)}^2$, and $\boldsymbol{D}_{(t+1)}$ are the current values of $\boldsymbol{\beta}$, σ^2, and $\boldsymbol{D}$, respectively.

M Step

The second step of the EM algorithm is the M step, which consists of updating the parameters by maximizing the conditional expected value of the complete likelihood. First, we can observe that for any value of $\boldsymbol{\theta}$, the value of $\boldsymbol{\beta}$ that maximizes $Q(\boldsymbol{\beta},\boldsymbol{\theta}|\boldsymbol{\beta}_{(t)},\boldsymbol{\theta}_{(t)})$ is given by

$$\boldsymbol{\beta}_{(t+1)} = \left(\boldsymbol{X}^{\mathrm{T}}\boldsymbol{X}\right)^{-1}\boldsymbol{X}^{\mathrm{T}}\left(\boldsymbol{y}-\boldsymbol{Z}\widetilde{\boldsymbol{b}}_{(t)}\right)$$

which does not depend on the chosen values of $\boldsymbol{\theta}$, but rather specifically on the chosen values of σ^2 and $\boldsymbol{D}$. Then by equating to zero, the derivative of $Q(\boldsymbol{\beta}_{(t+1)},\boldsymbol{\theta}|\boldsymbol{\beta}_{(t)},\boldsymbol{\theta}_{(t)})$ with respect to σ^2, and solving for σ^2, we can obtain that the value of σ^2 that maximizes $Q(\boldsymbol{\beta}_{(t+1)},\boldsymbol{\theta}|\boldsymbol{\beta}_{(t)},\boldsymbol{\theta}_{(t)})$, for $\boldsymbol{D}$ fixed, is given by

$$\sigma_{(t+1)}^2 = \frac{1}{n}\left[\operatorname{tr}\left(\boldsymbol{Z}\widetilde{\boldsymbol{D}}_{(t)}\boldsymbol{Z}^{\mathrm{T}}\right) + \left(\boldsymbol{y}-\boldsymbol{X}\boldsymbol{\beta}_{(t+1)}-\boldsymbol{Z}\widetilde{\boldsymbol{b}}_{(t)}\right)^{\mathrm{T}}\left(\boldsymbol{y}-\boldsymbol{X}\boldsymbol{\beta}_{(t+1)}-\boldsymbol{Z}\widetilde{\boldsymbol{b}}_{(t)}\right)\right]$$

which is independent of the value of $\boldsymbol{D}$. Now, according to result 4.10 in Johnson and Wichern (2002), the value of $\boldsymbol{D}$ that maximizes $Q(\boldsymbol{\beta},\boldsymbol{\theta}|\boldsymbol{\beta}_{(t)},\boldsymbol{\theta}_{(t)})$ is given by

$$\boldsymbol{D}_{(t+1)} = \widetilde{\boldsymbol{D}}_{(t)} + \widetilde{\boldsymbol{b}}_{(t)}\widetilde{\boldsymbol{b}}_{(t)}^{\mathrm{T}}$$

and does not depend on $\boldsymbol{\beta}$ and σ^2. So, by joining the above optimization, we have the M step that consists of updating the parameters $\boldsymbol{\beta}$, σ^2, and $\mathbf{D}$, with $\boldsymbol{\beta}_{(t+1)}$, $\sigma_{(t+1)}^2$, and $\boldsymbol{D}_{(t+1)}$, respectively. At this point, we can observe that the current value of the parameters $\boldsymbol{\beta}_{(t)}$, $\sigma_{(t)}^2$, and $\boldsymbol{D}_{(t)}$ are used in the computation of $\widetilde{\boldsymbol{b}}$ and $\widetilde{\boldsymbol{D}}$.

In the case of $\boldsymbol{D} = \sigma_g^2\boldsymbol{A}$, where $\boldsymbol{A}$ is a known matrix (the case in some genomic prediction models, where $\boldsymbol{A}$ corresponds to the genomic relationship matrix, the pedigree matrix, or the environmental matrix), the unique variance parameters to estimate are σ_g^2 and σ^2, that is, $\left(\boldsymbol{\theta}=\left(\sigma_g^2,\sigma^2\right)\right)$. In the same fashion, very often in genomic applications but also in a more general setting, $\boldsymbol{D}=\text{Diag}\left(\sigma_1^2\boldsymbol{A}_1,\ldots,\sigma_K^2\boldsymbol{A}_K\right)$, where $\boldsymbol{A}_k$ represents a known variance–covariance matrix (or correlation) structure (genomic relationship matrix, pedigree relationship matrix, etc., Burgueño et al. 2012) between the different random effects included. In this context, the variance component parameters to estimate are $\sigma_k^2, k=1,\ldots,K,$ and σ^2 $\left(\boldsymbol{\theta}=\left(\sigma_1^2,\ldots,\sigma_K^2,\sigma^2\right)\right)$, where $\boldsymbol{A}_k$, $k=1,\ldots,K$, are positive defined known matrices of dimensions $q_k\times q_k$, $k=1,\ldots,K$, such that $\sum_{k=1}^K q_k=q$, the E step can be reduced (see Appendix 2) to

$$Q\left(\boldsymbol{\beta},\boldsymbol{\theta}|\boldsymbol{\beta}_{(t)},\boldsymbol{\theta}_{(t)}\right)=-\frac{n}{2}\log\left(2\pi\sigma^2\right)-\frac{1}{2\sigma^2}\left[\text{tr}\left(\boldsymbol{Z}\widetilde{\boldsymbol{D}}_{(t)}\boldsymbol{Z}^{\text{T}}\right)+\left(\boldsymbol{y}-\boldsymbol{X}\boldsymbol{\beta}-\boldsymbol{Z}\widetilde{\boldsymbol{b}}_{(t)}\right)^{\text{T}}\left(\boldsymbol{y}-\boldsymbol{X}\boldsymbol{\beta}-\boldsymbol{Z}\widetilde{\boldsymbol{b}}_{(t)}\right)\right]$$
$$-\frac{1}{2}\sum_{k=1}^{K}\left\{\frac{\sigma_{k(t)}^2}{\sigma_k^2}\left[q_k-\sigma_{k(t)}^2\text{tr}\left(\boldsymbol{A}_k\boldsymbol{Z}_k^{\text{T}}\boldsymbol{V}_{(t)}^{-1}\boldsymbol{Z}_k\right)+\sigma_{k(t)}^{-2}\widetilde{\boldsymbol{b}}_{k(t)}^{\text{T}}\boldsymbol{A}_k^{-1}\widetilde{\boldsymbol{b}}_{k(t)}\right]+q_k\log\left(\sigma_k^2\right)+\log\left(|\boldsymbol{A}_k|\right)\right\},$$

where $\boldsymbol{V}_{(t)}$ is the marginal matrix of variance–covariance of the response vector in the current value of the parameters, $\mathbf{Z}=[\mathbf{Z}_1\mathbf{Z}_2\cdots\mathbf{Z}_K]$ is the partitioned design matrix of random effects, with $\mathbf{Z}_k$ $n\times q_k$, the corresponding matrix design for the random effects k, $\boldsymbol{b}_k$ $(\boldsymbol{b}^{\text{T}}=(\boldsymbol{b}_1^{\text{T}},\ldots,\boldsymbol{b}_K^{\text{T}}))$, $\widetilde{\boldsymbol{b}}_{(t)}^{\text{T}}=\left(\widetilde{\boldsymbol{b}}_{1(t)}^{\text{T}},\ldots,\widetilde{\boldsymbol{b}}_{K(t)}^{\text{T}}\right)$, and $\sigma_{k(t)}^2, k=1,\ldots,K,$ are the current values of the variance parameters. Finally, for this specific model, the maximization updates for the beta coefficients and variance components are the same as before, where the variance components are

$$\sigma_{k(t+1)}^2=\frac{1}{q_k}\sigma_{k(t)}^2\left[q_k-\sigma_{k(t)}^2\text{tr}\left(\boldsymbol{A}_k\boldsymbol{Z}_k^{\text{T}}\boldsymbol{V}_{(t)}^{-1}\boldsymbol{Z}_k\right)+\sigma_{k(t)}^{-2}\widetilde{\boldsymbol{b}}_{k(t)}^{\text{T}}\boldsymbol{A}_k^{-1}\widetilde{\boldsymbol{b}}_{k(t)}\right], k=1,\ldots,K.$$

These are obtained by maximizing $Q(\boldsymbol{\beta},\boldsymbol{\theta}|\boldsymbol{\beta}_{(t)},\boldsymbol{\theta}_{(t)})$ (defined above) with respect to $\sigma_k^2, k=1,\ldots,K$.

5.2.1.2 REML

An alternative to the ML estimation of the variance components of model (5.1) and to avoid the underestimation of the maximum likelihood method is the restricted maximum likelihood estimation method (REML) proposed by Patterson and Thompson (1971). Among the several ways to define this, one is discussed by Laird and Ware (1982), which under a Bayesian paradigm, consists of estimating the parameters of the variance components by maximizing the marginal posterior distribution of the variance components by assuming a "locally" uniform prior to the distribution for $\boldsymbol{\beta}$ and $\boldsymbol{\theta}$, that is, $f(\boldsymbol{\beta},\boldsymbol{\theta})\propto 1$ (Pinheiro and Bates 2000). The marginal posterior of the variance components is given in Appendix 3

$$f(\boldsymbol{\theta}|\boldsymbol{y}) \propto \int f(\boldsymbol{\beta},\boldsymbol{\theta}|\boldsymbol{y})d\boldsymbol{\beta} \propto \int f(\boldsymbol{y}|\boldsymbol{\beta},\boldsymbol{\theta})d\boldsymbol{\beta}$$
$$\propto |\boldsymbol{V}|^{-\frac{1}{2}}|\boldsymbol{X}^{\mathrm{T}}\boldsymbol{V}^{-1}\boldsymbol{X}|^{\frac{1}{2}}\exp\left\{-\frac{1}{2}\left(\boldsymbol{y}-\boldsymbol{X}\widetilde{\boldsymbol{\beta}}\right)^{\mathrm{T}}\boldsymbol{V}^{-1}\left(\boldsymbol{y}-\boldsymbol{X}\widetilde{\boldsymbol{\beta}}\right)\right\},$$

where $\widetilde{\boldsymbol{\beta}} = \left(\boldsymbol{X}^{\mathrm{T}}\boldsymbol{V}^{-1}\boldsymbol{X}\right)^{-1}\boldsymbol{X}^{\mathrm{T}}\boldsymbol{V}^{-1}\boldsymbol{y}$, which corresponds to the maximum likelihood of the fixed effects when the variance components are assumed known as generalized least squares estimates of $\boldsymbol{\beta}$ (GLS). Then, the restricted maximum likelihood estimators (REML) $\boldsymbol{\theta}$ are those that maximize $f(\boldsymbol{\theta}|\boldsymbol{y})$, or equivalently, the REML $\boldsymbol{\theta}$ can be defined as maximizing the

$$\ell_R(\boldsymbol{\theta};\boldsymbol{y}) = -\frac{1}{2}\log\left(|\boldsymbol{X}^{\mathrm{T}}\boldsymbol{V}^{-1}\boldsymbol{X}|\right) - \frac{1}{2}\log\left(|\boldsymbol{V}|\right) - \frac{1}{2}\left(\boldsymbol{y}-\boldsymbol{X}\widetilde{\boldsymbol{\beta}}\right)^{\mathrm{T}}\boldsymbol{V}^{-1}\left(\boldsymbol{y}-\boldsymbol{X}\widetilde{\boldsymbol{\beta}}\right)$$

This function is known as the restricted likelihood because it can be shown that this also corresponds to the likelihood associated with the maximum number $(n-p-1)$ of linearly independent error contrasts $\boldsymbol{FY}$, where $\boldsymbol{F}$ is a full row rank $(n-p-1)\times n$ known matrix such that $\boldsymbol{FX}=\boldsymbol{0}$. It is important to point out that the associated likelihood based on the transformed data, $\boldsymbol{FY}$, gives the same result for any chosen contrast error matrix $\boldsymbol{F}$ and, consequently, this is invariant to fixed effect parameters (Harville 1974). Equivalently, the REML of $\boldsymbol{\beta}$ and $\boldsymbol{\theta}$ can be defined as those that maximize the

$$\ell_R(\boldsymbol{\beta},\boldsymbol{\theta};\boldsymbol{y}) = -\frac{1}{2}\log\left(|\boldsymbol{X}^{\mathrm{T}}\boldsymbol{V}^{-1}\boldsymbol{X}|\right) - \frac{1}{2}\log\left(|\boldsymbol{V}|\right) - \frac{1}{2}(\boldsymbol{y}-\boldsymbol{X}\boldsymbol{\beta})^{\mathrm{T}}\boldsymbol{V}^{-1}(\boldsymbol{y}-\boldsymbol{X}\boldsymbol{\beta})$$

This objective function is like the natural logarithm of likelihood function given in Eq. (5.2) (log-likelihood) except for the first term. To obtain the REML solutions or the maximum a posteriory (when adopting a locally uniform prior for the parameter, as described before) of the variance components parameters, (like for the MLE) numerical methods are required. See Jennrich and Schluchter (1986) and Lindstrom and Bates (1988) for details on the Newton–Raphson and Fisher Scoring algorithms; consult the lme4 R package (Bates et al. 2015) which uses a generic nonlinear optimizer and implements a large variety of different models (MLE and REML) that arise from the LMM when different structures of the variance of random effects and errors are adopted. For a derivation of the EM algorithm to obtain the REML, see Laird and Ware (1982) under model (5.1) for longitudinal data; this same approach can be used for the genomic model previously described, where $\boldsymbol{D} = \mathrm{Diag}\left(\sigma_1^2\boldsymbol{A}_1,\ldots,\sigma_K^2\boldsymbol{A}_K\right)$ and $\boldsymbol{R} = \sigma^2\boldsymbol{I}_n$. Consult Searle (1993) and Covarrubias-Pazaran (2016, 2018) for an implementation of this algorithm with the EM function in the sommer R package.

5.2.1.3 BLUPs

In many situations, in addition to the estimation of the fixed effects, the prediction of the random effects is also of interest. A standard method for "estimating" the random effects is the best linear unbiased predictor (BLUP; Robinson 1991), which originally was developed by Henderson (1975) in animal breeding for estimating merit in dairy cattle and is now commonly employed in many research areas (Piepho et al. 2008). If the variance components $\boldsymbol{D}$ and $\boldsymbol{R}$ are known, the best linear unbiased predictor (BLUP) of the random effects $\boldsymbol{b}$ is given by

$$\widetilde{\boldsymbol{b}}^* = \boldsymbol{D}\boldsymbol{Z}^{\mathrm{T}}\boldsymbol{V}^{-1}\left(\boldsymbol{y} - \boldsymbol{X}\widetilde{\boldsymbol{\beta}}\right),$$

where $\widetilde{\boldsymbol{\beta}} = \left(\boldsymbol{X}^{\mathrm{T}}\boldsymbol{V}^{-1}\boldsymbol{X}\right)^{-1}\boldsymbol{X}^{\mathrm{T}}\boldsymbol{V}^{-1}\boldsymbol{y}$ is the generalized least squared (GLS) estimator of $\boldsymbol{\beta}$. This can be obtained by maximizing with respect to $\boldsymbol{\beta}$ and $\boldsymbol{b}$, the joint density of $\boldsymbol{y}$ and $\boldsymbol{b}$, $f_{Y,\boldsymbol{b}}(\boldsymbol{y},\boldsymbol{b})$, and is the reason why Harville (1985) called these estimates of realized values $\boldsymbol{b}$ (McLean et al. 1991), or likewise by solving the mixed model equations (MME) (Henderson 1975):

$$\begin{bmatrix} \boldsymbol{X}^{\mathrm{T}}\boldsymbol{R}^{-1}\boldsymbol{X} & \boldsymbol{X}^{\mathrm{T}}\boldsymbol{R}^{-1}\boldsymbol{Z} \\ \boldsymbol{Z}^{\mathrm{T}}\boldsymbol{R}^{-1}\boldsymbol{X} & \boldsymbol{Z}^{\mathrm{T}}\boldsymbol{R}^{-1}\boldsymbol{Z}+\boldsymbol{D}^{-1} \end{bmatrix} \begin{bmatrix} \widetilde{\boldsymbol{\beta}} \\ \widetilde{\boldsymbol{b}}^* \end{bmatrix} = \begin{bmatrix} \boldsymbol{X}^{\mathrm{T}}\boldsymbol{R}^{-1}\boldsymbol{y} \\ \boldsymbol{Z}^{\mathrm{T}}\boldsymbol{R}^{-1}\boldsymbol{y} \end{bmatrix}$$

from which the inversion of the variance–covariance matrix of $\boldsymbol{Y}$ is avoided, which can be helpful in some situations to save considerable computational resources. Note that $\widetilde{\boldsymbol{b}}^*$ corresponds to the posterior mean of the random effects where the fixed effects are replaced by its GLS (Searle et al. 2006).

When the variance components are unknown, which is most often the case, they are frequently estimated using restricted maximum likelihood estimators, which replace them in the corresponding equations. Then, the approximate best linear unbiased predictor is obtained and is referred to as the estimated or empirical best linear unbiased predictor (EBLUP) (Rencher 2008).

To solve the mixed model equations, there are several software packages that can be useful, but one in particular in the genomic context is the sommer package (Covarrubias-Pazaran 2016, 2018) that internally solves the MME after the variance components are estimated. The github version of the sommer R package can be accessed at https://github.com/cran/sommer and can be installed with the following commands:

```
install.packages('devtools');
library(devtools);
install_github('covaruber/sommer')
```

5.3 Linear Mixed Models in Genomic Prediction

In a simple genomic prediction context where $\boldsymbol{b}$ includes the genotype effects, and the genomic relationship matrix (VanRaden 2008), that is, $\boldsymbol{G}$ is available, very often the assumed variance–covariance matrix of the random effects is $\sigma_g^2\boldsymbol{G}$ and the errors are assumed as independently and identically distributed, $\boldsymbol{R} = \sigma^2\boldsymbol{I}_n$, where n is the total number of observations. In this case, the resultant model is known as the GBLUP model and when the pedigree is used, it is referred to as PBLUP. Another kind of information between lines can also be used, such as the relationship matrices derived from hyperspectral reflectance information (Krause et al. 2019). Other extensions of this model can be developed by taking into account other factors, for example, genotype$\times$ environment interaction, as will be illustrated later in the genomic prediction context.

In this case, where only the genotypic effects are taken into account, in the linear mixed model (5.1), the fixed effects design matrix is $\boldsymbol{X} = \mathbf{1}_n$, where the vector of length n corresponds to the general mean $\boldsymbol{\beta} = \beta_0$, $\boldsymbol{b} = (b_1, b_2, \ldots, b_J)^{\mathrm{T}}$ contains the genotypic effects of J lines, and $\boldsymbol{Z}$ is the incidence matrix design for the random line effects ($\boldsymbol{Z}_L$):

$$Y = \mathbf{1}_n\mu + Z_L\boldsymbol{b} + \boldsymbol{\epsilon}, \tag{5.3}$$

where $\boldsymbol{b} \sim N_J\left(\mathbf{0}, \sigma_g^2\boldsymbol{G}\right)$ and $\boldsymbol{R} = \sigma^2\boldsymbol{I}_n$.

The basic code to implement the GBLUP model (5.3) with the sommer package is the following:

```
A = mmer(y ~ 1,random= ~ vs(GID,Gu=G) , rcov= ~ vs(units),
         data=dat_F, verbose=FALSE)
```

where **y** and **GID** are the column names that contain the response variable and genotypes in data set **dat_F**. **G** is the genomic relationship matrix for lines which is specified in the Gu argument, that in general serves to provide a known variance–covariance matrix between the levels of the random effects (GID). In the "rcov" option, the argument "units" is always used to specify the error term.

5.4 Illustrative Examples of the Univariate LMM

Example 1 To illustrate the performance of the LMM in a genomic prediction context doing the fitting process with the sommer package, we considered a wheat data set that consisted of 500 markers measured for each line as the genomic information, and with 229 observations in total that registered grain yield (tons/ha): 30 lines in four environments with one or two repetitions.

Table 5.1 Prediction performance of the GBLUP model (5.3, M1) and the model (5.3) that results from ignoring the genomic information (M10): mean squared error of prediction (MSE) and Pearson's correlation (PC), and its standard deviation for each criterion is reported for each partition

	M1		M10	
PT	MSE	PC	MSE	PC
1	0.7025	0.6917	0.6924	0.6969
2	0.4583	0.6681	0.4407	0.7044
3	0.5075	0.5251	0.4923	0.5506
4	0.4719	0.6024	0.4468	0.6328
5	0.6433	0.5479	0.6433	0.5307
6	0.3657	0.5088	0.3569	0.5255
7	0.6923	0.5081	0.6777	0.5243
8	0.3285	0.5054	0.2952	0.57
9	0.4364	0.6855	0.429	0.6962
10	0.6807	0.6374	0.6513	0.6811
Average (SD)	0.5287 (0.14)	0.5881 (0.078)	0.5126 (0.143)	0.6112 (0.078)

The prediction performance of the model given in Eq. (5.3) (**M1**) was evaluated with 10 random partitions, where each partition was made up of two subsets, one containing 80% of the data and used for training the model, and the other containing the remaining 20% of the data and used to evaluate the prediction performance of the model in terms of the mean squared error of prediction (MSE).

Furthermore, this model assumes that the errors were independently and identically distributed as $\boldsymbol{\epsilon} \sim N_n(\mathbf{0}, \sigma^2 \boldsymbol{I}_n)$; independently of the genotypic effects, $\boldsymbol{b}$ was assumed multivariate normal with a null mean vector and a variance–covariance matrix equal to $\sigma_g^2 \boldsymbol{G}$, where the genomic relationship matrix was computed with the information of the 500 markers.

The variance components parameters, in this case σ_g^2 and σ^2, were estimated by restricted maximum likelihood estimation with the mmer function in the sommer package, using the default algorithm optimization, the Newton–Raphson method. For univariate response variables, the EM algorithm through the EM function in this R package can also be used.

The results are shown in Table 5.1, where we also present the Pearson's correlation (PC) and MSE of the same model but without taking into account the information of the genomic relationship between lines ($\boldsymbol{G}$), that is, the variance–covariance matrix for the genotypic effects is assumed to be $\text{Var}(\boldsymbol{b}) = \sigma_g^2 \boldsymbol{I}_J$. This model is referred to as **M10**. From this table, we can observe that model **M10** shows a slightly better performance in terms of both MSE and PC criteria than the **M1** model: the MSE of model **M1** is 3.15% greater than the MSE of **M10**, while the PC of the **M10** is 3.94% greater than the corresponding **M1** model. The better average performance was observed with model **M10**, which did not consider genomic information, suggests that the marker information in this particular case did not provide useful information; however, in general, this is not expected when using marker information for prediction, although this could change with larger data sets (more lines and more markers) or by improving the quality of the available data.

The R code to reproduce this result is given in Appendix 4. This can be adapted easily to another CV strategy of interest where the objective, for example, can be the prediction of non-observed lines in some environments or the prediction of lines in a future year.

An extension of the GBLUP model is the G×E BLUP model that takes into account the main environmental effects, the genotypic effects, and the genotype ×environment interaction effects:

$$Y = \mathbf{1}_n\mu + \boldsymbol{X}_E\boldsymbol{\beta}_E + \boldsymbol{Z}_L\boldsymbol{b}_1 + \boldsymbol{Z}_{EL}\boldsymbol{b}_2 + \boldsymbol{\epsilon} \tag{5.4}$$

where now the fixed effects are part of the linear mixed model (5.1) that was explicitly split into the general mean part ($\mathbf{1}_n\mu$) and the environment effects term ($\boldsymbol{X}_E\boldsymbol{\beta}_E$), $\boldsymbol{X} = [\mathbf{1}_n\, \boldsymbol{X}_E]$ and $\boldsymbol{\beta} = \left(\mu, \boldsymbol{\beta}_E^{\mathrm{T}}\right)^{\mathrm{T}}$. Similarly, for the random effects, $\boldsymbol{Z} = [\boldsymbol{Z}_L\, \boldsymbol{Z}_{EL}]$ and $\boldsymbol{b} = \left[\boldsymbol{b}_1^{\mathrm{T}}, \boldsymbol{b}_2^{\mathrm{T}}\right]^{\mathrm{T}}$, where $\boldsymbol{b}_1$ and $\boldsymbol{b}_2$ were the vectors with the random genotypic effects and the vector with the random genotype×environment interaction effects, with incidence matrix $\boldsymbol{Z}_L$ and $\boldsymbol{Z}_{EL}$, respectively. For $\boldsymbol{b}_1$, the same distribution as the GBLUP model was assumed, $\boldsymbol{b}_1 \sim N_J\left(\mathbf{0}, \sigma_g^2\boldsymbol{G}\right)$, and for the second random effect, $\boldsymbol{b}_2 \sim N_J(\mathbf{0}, \boldsymbol{\Sigma}_E \otimes \boldsymbol{G})$, where $\boldsymbol{\Sigma}_E \otimes \boldsymbol{G}$ is the relationship matrix of the genotype×environment interaction term, with $\boldsymbol{\Sigma}_E$ the genetic variance–covariance matrix between I environments; the ith element of the diagonal of $\boldsymbol{\Sigma}_E$, σ_{Ei}^2, is the genetic variance in environment i, $i = 1, \ldots, I$, and $\sigma_{Eik}\boldsymbol{G}$ is the genetic variance–covariance matrix for lines in environments i and k, where σ_{Eik} is the element (i, k) of $\boldsymbol{\Sigma}_E$.

When $\boldsymbol{\Sigma}_E$ has a non-diagonal structure, the information from the genomic relationship matrix and the correlated environments can be helpful for improving the prediction performance of the model by borrowing information between lines inside an environment and between lines across and among environments (Burgueño et al. 2012).

Example 2 To illustrate how model (5.4) can be implemented using the sommer package, the same data used in Example 1 are considered, where the same 30 genotypes are in the four environments. Besides the line indicator (GID), environment information (Env) was also available in the data set, which was needed for implementing model (5.4). The adopted structure for the variance–covariance matrix between environments is $\boldsymbol{\Sigma}_E = \sigma_{EG}^2\boldsymbol{I}_I$ and the resulting model is referred to as M2. Another explored model (M20) was obtained under the same specification, with the difference that $\boldsymbol{G}$ was set equal to the identity matrix.

Using the same validation scheme that was used in Example 1, the results for each of the 10 random partitions are shown in Table 5.2, in which, for illustrative purposes, model (5.3) plus environment as a fixed effect (M11) is also included, that is,

Table 5.2 Prediction performance of two sub-models of (5.4): model M2 in which $\text{Var}(\boldsymbol{b}_1) = \sigma_g^2 \boldsymbol{G}$, $\boldsymbol{b}_2 \sim N_J(\boldsymbol{0}, \boldsymbol{\Sigma}_E \otimes \boldsymbol{G})$ and $\boldsymbol{\Sigma}_E = \sigma_{EG}^2 \boldsymbol{I}_I$ (M2); and model M20 that is the same as model M2 but the genomic information is not taken into account, that is, $\boldsymbol{G} = \boldsymbol{I}_J$

	M2		M20		M11	
PT	MSE	PC	MSE	PC	MSE	PC
1	0.5641	0.8198	0.5364	0.8576	0.5658	0.8168
2	0.4292	0.7645	0.4014	0.8393	0.4447	0.6424
3	0.3865	0.7873	0.3621	0.8497	0.4246	0.6422
4	0.45	0.6997	0.401	0.7897	0.3853	0.6992
5	0.6879	0.5307	0.6636	0.5902	0.595	0.5933
6	0.3048	0.6802	0.2819	0.7326	0.3267	0.5777
7	0.6405	0.6823	0.6369	0.7066	0.558	0.6914
8	0.4132	0.5261	0.4053	0.5562	0.2999	0.5817
9	0.3217	0.8384	0.2978	0.8869	0.3457	0.7797
10	0.5397	0.8368	0.5206	0.8572	0.5647	0.7307
Average (SD)	0.4738 (0.13)	0.7166 (0.116)	0.4507 (0.133)	0.7666 (0.117)	0.451 (0.112)	0.6755 (0.083)

M11 is referred to as model (5.3) plus environment effects (Env). The mean squared error of prediction (MSE) and Pearson's correlation (PC) for each partition are reported. SD is the standard deviation

$$\boldsymbol{Y} = \boldsymbol{1}_n \mu + \boldsymbol{X}_E \boldsymbol{\beta}_E + \boldsymbol{Z}_L \boldsymbol{b} + \boldsymbol{\epsilon},$$

where μ, $\boldsymbol{\beta}_E$, and $\boldsymbol{b}$ are as before (5.3), and $\boldsymbol{X}_E \boldsymbol{\beta}_E$ is the predictor term corresponding to the environment fixed effects.

From Table 5.2 we can observe yet again a moderately better performance of model M20 that does not take into account the genomic information. Model M2 also confirms the lack of usefulness of the marker information in this case, but again, this in general is expected to change for other data sets with a greater number of lines, markers, or more data quality. The MSE of model M2 is 5.11% greater than the MSE of model M20, while the PC value of this last model is 6.98% greater than the corresponding PC value obtained with the M20 model. When comparing the M20 model and M11, the MSE of this last model is just 0.075% greater than the corresponding MSE of M20, but when considering the PC value, the first model resulted in 13.98% greater than model (5.1) plus the environment effect. Indeed, because of the high variation observed across partitions (SD values of PC and MSE), there is no significant difference between models in Table 5.2.

Furthermore, in terms of the average MSE, the model with the best performance between those presented in Table 5.1 (M10) is 13.73% greater than the average MSE of the best performance model between those compared in Table 5.2 (M11), while in terms of the average Pearson's correlation, the best model in Table 5.2 (M20) is 25.42% greater than the average Pearson's correlation of the best model in Table 5.1. The worse average MSE performance of those in Table 5.1 (M1) is 17.31% greater than the best average MSE performance in Table 5.2 (M20), and the best average PC performance in Table 5.2 (M20) is 30.37% greater than the worse average PC performance in Table 5.1 (M1). Actually, the best average MSE model in Table 5.1 (M10) is 8.19 greater than the worse average MSE model in Table 5.2 (M2), while the worse average PC model in Table 5.2 (M11) is 10.51% greater than the average PC model in Table 5.1 (M10).

The R code to reproduce these results is given in Appendix 5.

Other versions of model (5.4) can be obtained by adopting other variance–covariance structures. For example, another version of model (5.4) can be obtained when environment covariates are available (***W***) and they are used to model the $G\times E$ predictor term, specifically when the genetic variance–covariance matrix between environments, $\mathbf{\Sigma}_E$, is modeled by $\mathbf{\Sigma}_E = \sigma^2_{EG}\boldsymbol{O}$, where $\boldsymbol{O} = \frac{1}{p_w}\boldsymbol{W}\boldsymbol{W}^{\mathrm{T}}$ and the similarity between environments is computed like the genomic relationship matrix (**G**), using the information of p_w environment covariates (Jarquín et al. 2014; Martini et al. 2020), or where $\boldsymbol{O}$ is obtained from phenotypic correlations across environments from related historical data (Martini et al. 2020). In the sommer package, this can be implemented using the following basic R code:

```
O = diag(I) #Specified the O matrix for the I environments
dat_F$Env_GID = paste(dat_F$Env,dat_F$GID,sep='_')
GE = kronecker(O,G)
rnGWE = expand.grid(row.names(G),unique(dat_F$Env))
row.names(GE) = paste(rnGWE[,2],rnGWE[,1],sep='_')
colnames(GE) = row.names(GE)
A = mmer(y ~ Env, random= ~ vs(GID,Gu=G) + vs(Env_GID,Gu = GE),
     rcov= ~ vs(units), data=dat_F)
```

Other more complex models can be explored with the sommer package when more data information is available, such as specifying an unstructured variance–covariance matrix for $\mathbf{\Sigma}_E$. A simpler model is the non-correlated heterogeneous variance components (for environments) which arises by assuming a diagonal structure, $\mathbf{\Sigma}_E = \mathrm{Diag}\left(\sigma_1^2, \ldots, \sigma_I^2\right)$. This can be implemented by replacing the interaction term in the predictor in sommer *vs(**Env**,Gu* = ***G****)* by *vs(ds(**Env**), **GID**,Gu* = ***G****)*. Similarly, for a specific environment residual variance, *vs(units)* need to be replaced by *vs(ds(Env),units)* or *vs(at(Env),units)*. See Appendix 7 for a basic code to implement all these models and see Covarrubias-Pazaran (2018) for more variance structures that can be exploited in this model.

5.5 Multi-trait Genomic Linear Mixed-Effects Models

In some genomic applications, there are several traits of interest and all of them are measured in some lines but in other lines only subsets of those traits are measured. Although separate univariate genomic linear mixed models can be performed to analyze all measured traits, sometimes single univariate genomic models do not work well, especially in traits with low heritability. When low heritability traits have at least moderate correlation with high heritability traits, the prediction performance ability for these low heritability traits could strongly increase by using a multi-trait model (Jia and Jannink 2012; Montesinos-López et al. 2016; Budhlakoti et al. 2019).

If for each line ($j = 1, \ldots J$), n_T traits are measured, Y_{jt}, $t = 1, \ldots n_T$, the multi-trait genomic linear mixed-effects model adopts an unstructured covariance matrix for the residuals between traits and for the random genotypic effects between traits, and similar to the univariate trait models (5.3), this can be expressed as

$$\begin{bmatrix} Y_{j1} \\ Y_{j2} \\ \vdots \\ Y_{jn_T} \end{bmatrix} = \begin{bmatrix} \mu_1 \\ \mu_2 \\ \vdots \\ \mu_{n_T} \end{bmatrix} + \begin{bmatrix} g_{j1} \\ g_{j2} \\ \vdots \\ g_{jn_T} \end{bmatrix} + \begin{bmatrix} \epsilon_{j1} \\ \epsilon_{j2} \\ \vdots \\ \epsilon_{jn_T} \end{bmatrix}, j = 1, \ldots, J, \tag{5.5}$$

where μ_t, $t = 1, \ldots, n_T$, are the specific trait means, g_{jt}, $t = 1, \ldots, n_T$, are the specific trait genotypic effects, and ϵ_{jt}, $t = 1, \ldots, n_T$, are the random error terms corresponding to each trait. Furthermore, $\boldsymbol{b} = [\boldsymbol{g}_1^{\mathrm{T}}, \ldots, \boldsymbol{g}_J^{\mathrm{T}}]^{\mathrm{T}} \sim N(\boldsymbol{0}, \boldsymbol{G}\otimes\boldsymbol{\Sigma}_{\mathrm{T}})$, $\boldsymbol{g}_j = [g_{j1}, \ldots, g_{jn_T}]^{\mathrm{T}}$, $j = 1, \ldots, J$, and $\boldsymbol{\epsilon}_j = [\epsilon_{j1}, \ldots, \epsilon_{jn_T}]^{\mathrm{T}}$, $j = 1, \ldots, J$, are independent multivariate normal random vectors with null mean and variance $\mathbf{R}_{n_T}$, $\boldsymbol{\Sigma}_T$ is $n_T \times n_T$ matrix that represents the genetic covariance between traits, and $\otimes$ is the Kronecker product.

In matrix notation, it is the linear mixed model (5.1) where $\boldsymbol{Y} = [\boldsymbol{Y}_1^{\mathrm{T}}, \ldots, \boldsymbol{Y}_J^{\mathrm{T}}]^{\mathrm{T}}$, $\boldsymbol{Y}_j = [Y_{j1}, \ldots, Y_{jn_T}]^{\mathrm{T}}$, $\boldsymbol{X} = \boldsymbol{1}_J\otimes\boldsymbol{I}_{\mathrm{n_T}}$, $\boldsymbol{\beta} = \boldsymbol{\mu} = (\mu_1, \ldots, \mu_{n_T})^{\mathrm{T}}$, $\boldsymbol{Z} = \boldsymbol{I}_{n_T J}$, $\boldsymbol{b} = [\boldsymbol{g}_1^{\mathrm{T}}, \ldots, \boldsymbol{g}_J^{\mathrm{T}}]^{\mathrm{T}}, \boldsymbol{\epsilon} = [\boldsymbol{\epsilon}_1^{\mathrm{T}} \ldots \boldsymbol{\epsilon}_J^{\mathrm{T}}]^{\mathrm{T}} \sim N(\boldsymbol{0}, \boldsymbol{I}_J\otimes\boldsymbol{R}_{\mathrm{n_T}})$, and $\boldsymbol{b} = [\boldsymbol{g}_1^{\mathrm{T}}, \ldots, \boldsymbol{g}_J^{\mathrm{T}}]^{\mathrm{T}} \sim N(\boldsymbol{0}, \boldsymbol{G}\otimes\boldsymbol{\Sigma}_{\mathrm{T}})$. Similarly, the extended model that arises by adding more fixed effects ($\boldsymbol{X}$) can be specified by adding a term $\boldsymbol{X\beta}$ to the predictor:

$$\boldsymbol{Y} = (\boldsymbol{1}_{IJ}\otimes\boldsymbol{I}_{\mathrm{n_T}})\boldsymbol{\mu} + \boldsymbol{X\beta} + \boldsymbol{Zb} + \boldsymbol{\epsilon} \tag{5.5a}$$

When $\boldsymbol{\Sigma}_T$ and $\boldsymbol{R}$ are diagonal matrices, model (5.5) is equivalent to separately fitting a univariate GBLUP model to each trait.

The R code to fit this multivariate model with the sommer package is

```
A = mmer(cbind(T1,...,TnT) ~x1+x2+...+xp ,
        random= ~ vs(GID,Gu=G), rcov= ~ vs(units), data=dat_F)
```

where **y** and **GID** are again the column names corresponding to the response variables and genotypes in data set **dat_F**, while *T1*, …, *TnT* are the column names of the matrix of response variables (**y**) in **dat_F** corresponding to the traits to be used, and similarly, *x1*, *x2*,…, *xp* are the column names of *p* covariates to be included in the fitting process (see below the R code for Example 3). The rest of the arguments are the same as those described in the R code of model 5.3.

Example 3 To illustrate the fitting of the multi-trait genomic model (5.5) (M3), we considered the same data set used in Examples 1 and 2, but with the addition of trait (y2) to be able to explore the implementation of a bivariate trait genomic model. The same CV strategy implemented in Example 1 was used. In addition to model (5.5), we also evaluated a sub-model that was obtained by considering a diagonal structure for $\mathbf{\Sigma}_T$, that is, $\mathbf{\Sigma}_T = \text{Diag}\left(\sigma_{T_1}^2, \sigma_{T_2}^2\right)$. This model will be referred to as M32.

The results are in Table 5.3. On average, the two evaluated models (M3 and M32) showed a similar performance in terms of the two criteria used, MSE and PC, for both traits, but in all partitions a slightly better performance was observed in favor of model M3. For trait T1, the simpler model (M32) gave an MSE 0.785% greater than model M3, while the more complex model (M3) gave a PC only 1.066% greater than that of model M32. The difference was less for the second trait (T2), where the average MSE of M32 was only 0.165% greater than the one corresponding to model M3, while the PC of M3 was only 0.046% greater than the PC of M32.

Furthermore, note that the difference between the univariate models presented in Tables 5.1 and 5.2 and the multivariate models of Table 5.3 is not significant (only on average the models in Table 5.2 result better than models in Table 5.1) because the large standard deviation observed across partitions in MSE and PC, which in this case indicate that the multivariate model does not help improve the prediction accuracy in the trait of interest (first trait). But as commented before, this benefit could be obtained with more related auxiliary secondary traits and larger data sets of good quality.

Table 5.3 Prediction performance of a bivariate trait model (5.5)

Models	M3				M32			
Trait	T1		T2		T1		T2	
PT	MSE	PC	MSE	PC	MSE	PC	MSE	PC
1	0.6959	0.6954	0.0978	0.984	0.6973	0.6939	0.0982	0.9835
2	0.4476	0.6851	0.1411	0.9256	0.4526	0.6759	0.1409	0.9253
3	0.494	0.5448	0.0446	0.9846	0.503	0.5315	0.0436	0.985
4	0.4636	0.6125	0.1258	0.9665	0.4674	0.6079	0.1266	0.9663
5	0.6398	0.5496	0.1588	0.9306	0.6413	0.5489	0.1596	0.93
6	0.3598	0.5192	0.0691	0.9775	0.3632	0.5134	0.0693	0.9772
7	0.6864	0.514	0.0273	0.9953	0.691	0.5094	0.0272	0.9951
8	0.3104	0.5421	0.2011	0.9492	0.3187	0.5267	0.2	0.9482
9	0.4394	0.678	0.1983	0.9468	0.4359	0.6833	0.2005	0.9452
10	0.6685	0.6575	0.0838	0.974	0.6759	0.644	0.0837	0.9738
Average (SD)	0.5205 (0.142)	0.5998 (0.074)	0.1148 (0.061)	0.9634 (0.024)	0.5246 (0.141)	0.5935 (0.076)	0.115 (0.061)	0.963 (0.024)

This model is referred to as M3 and when assuming a diagonal structure for $\mathbf{\Sigma}_T$, $\mathbf{\Sigma}_T = \text{Diag}\left(\sigma_{1T}^2, \sigma_{2T}^2\right)$, it is referred to as M32. The mean squared error of prediction (MSE) and Pearson's correlation (PC) for each trait in each partition are reported. SD is the standard deviation

The R code to obtain the results given in Table 5.3 is provided in Appendix 6. At the end of this Appendix, in the comment lines, the code is also available for a CV strategy, when we are interested in evaluating the performance of a bivariate model where only trait y2 is missing in testing data set and all the information of the other trait (y1) is available. This could be useful in real applications where the interest lies in predicting traits that are difficult or expensive to measure, with the phenotypic information of correlated traits that are easy or inexpensive to measure (Calus and Veerkamp 2011; Jiang et al. 2015). Of course, the code could be adapted for any other relevant strategy.

In a similar fashion, just as univariate genomic linear mixed model (5.4), model (5.5) can be directly extended to a model that considers the genotype ×environment interaction term. Next, we do this for the balanced case, and for this we assume that for each environment $i = 1, \ldots, I$, J lines were phenotyped for n_T traits, Y_{ijt}, $t = 1, \ldots, n_T$. In matrix notation, the extended $G\times E$ model (5.4) plus fixed effects ($\boldsymbol{X\beta}$) is given by

$$\boldsymbol{Y} = (\boldsymbol{1}_{IJ} \otimes \boldsymbol{I}_{n_T})\boldsymbol{\mu} + \boldsymbol{X\beta} + \boldsymbol{Z}_L\boldsymbol{b}_1 + \boldsymbol{Z}_{EL}\boldsymbol{b}_2 + \boldsymbol{\epsilon}, \tag{5.6}$$

where $\boldsymbol{Y} = \left[\boldsymbol{Y}_1^{\mathrm{T}} \ldots \boldsymbol{Y}_I^{\mathrm{T}}\right]^{\mathrm{T}}$, $\boldsymbol{Y}_i = \left[\boldsymbol{Y}_{i1}^{\mathrm{T}}, \ldots, \boldsymbol{Y}_{iJ}^{\mathrm{T}}\right]^{\mathrm{T}}$, $\boldsymbol{Y}_{ij} = \left[\boldsymbol{Y}_{ij1}, \ldots \boldsymbol{Y}_{ijn_T}\right]^{\mathrm{T}}$, $i = 1, \ldots, I$, $j = 1, \ldots, J$, $\boldsymbol{1}_{IJ}$ is the vector of ones of order IJ, $\boldsymbol{I}_{n_T}$ is the identity matrix of dimension n_T, $\boldsymbol{\mu} = \left(\mu_1, \ldots, \mu_{n_T}\right)^{\mathrm{T}}$ is the vector with the general specific trait means, $\boldsymbol{Z}_L = \boldsymbol{1}_I \otimes \boldsymbol{I}_{n_T J}$ and $\boldsymbol{Z}_{EL} = \boldsymbol{I}_{IJn_T}$ are the incidence matrices of genotype random effects ($\boldsymbol{b}_1$) and the incidence matrices of the genotype×environment interactions random effects ($\boldsymbol{b}_2$), respectively, with $\boldsymbol{b}_1 = \left[\boldsymbol{g}_1^{\mathrm{T}}, \ldots, \boldsymbol{g}_J^{\mathrm{T}}\right]^{\mathrm{T}}$ and $\boldsymbol{b}_2 = \left[\boldsymbol{g}_{21}^{\mathrm{T}}, \ldots, \boldsymbol{g}_{2I}^{\mathrm{T}}\right]^{\mathrm{T}}$, $\boldsymbol{g}_j = \left[g_{j1}, \ldots, g_{jn_T}\right]^{\mathrm{T}}$, $\boldsymbol{g}_{2i} = \left[\boldsymbol{g}_{2i1}^{\mathrm{T}}, \ldots, \boldsymbol{g}_{2iJ}^{\mathrm{T}}\right]^{\mathrm{T}}$, and $\boldsymbol{g}_{2ij} = \left[g_{2ij1}, \ldots g_{2ijn_T}\right]^{\mathrm{T}}$, $i = 1, \ldots, I$, $j = 1, \ldots, J$. In addition, it is assumed that $\boldsymbol{\epsilon} = \left[\boldsymbol{\epsilon}_1^{\mathrm{T}} \ldots \boldsymbol{\epsilon}_I^{\mathrm{T}}\right]^{\mathrm{T}} \sim N(\boldsymbol{0}, \boldsymbol{I}_{IJ} \otimes \boldsymbol{R}_{n_T})$, $\boldsymbol{b}_1 \sim N(\boldsymbol{0}, \boldsymbol{G} \otimes \boldsymbol{\Sigma}_T)$, and $\boldsymbol{b}_2 \sim N(\boldsymbol{0}, \boldsymbol{\Sigma}_E \otimes \boldsymbol{G} \otimes \boldsymbol{\Sigma}_{2T})$.

This shows that when $\boldsymbol{\Sigma}_T$, $\boldsymbol{\Sigma}_{2T}$, $\boldsymbol{\Sigma}_E$, and $\boldsymbol{R}$ are diagonal matrices, model (5.6) is equivalent to separately fitting a univariate GBLUP model for each trait.

Example 4 To illustrate the fitting and evaluation process of model (5.6), we considered a data set that contains the information of two traits, for which 150 lines were phenotyped each in two environments, and given a total of 300 bivariate phenotypic data points. Also, a genomic relationship matrix for the lines is available that was computed with marker information.

The first explored model is referred to as M4 and assumes an unstructured variance–covariance matrix for all the components in model (5.6), except for the assumption that $\boldsymbol{\Sigma}_E = \boldsymbol{I}_I$, i.e., the model assumes the same variance–covariance among environments. In addition to this model (M4), three sub-models were also explored: M42, which considers a diagonal structure for the genetic variance–covariance between traits, $\boldsymbol{\Sigma}_{2T} = \mathrm{Diag}\left(\sigma_{1T_1}^2, \sigma_{1T_2}^2\right)$, model M43 in which $\boldsymbol{\Sigma}_{1T} = \mathrm{Diag}\left(\sigma_{1T_1}^2, \sigma_{1T_2}^2\right)$ and $\boldsymbol{\Sigma}_{2T} = \mathrm{Diag}\left(\sigma_{2T_1}^2, \sigma_{2T_2}^2\right)$, and model M44 which is the same as

M43 but with an assumed diagonal variance–covariance matrix structure for the error, $\boldsymbol{R} = \text{Diag}\left(\sigma_{e_1}^2, \sigma_{e_2}^2\right)$.

The results are shown in Table 5.4, from which we can observe that for trait 1 (GY), the best performance under both criteria (MSE and PC) was obtained with the more complex model: M4. For this trait (GY), the MSE of models M42, M43, and M44 were 7.53%, 8.21%, and 8.17%, respectively, greater than the MSE of

Table 5.4 Prediction performance of some sub-models of model (5.6)

Model	M4				M42			
Trait	T1 (GY)		T2 (Testwt)		T1 (GY)		T2 (Testwt)	
PT	MSE	PC	MSE	PC	MSE	PC	MSE	PC
1	0.1895	0.5962	1.0291	0.6416	0.1717	0.5691	0.9357	0.6942
2	0.2414	0.5091	0.8139	0.6466	0.2444	0.5127	0.7199	0.6965
3	0.2638	0.8089	1.2914	0.5875	0.2324	0.5661	1.0839	0.7705
4	0.475	0.1266	0.9975	0.7118	0.4856	0.101	1.0271	0.6846
5	0.2728	0.2237	0.8053	0.6897	0.272	0.2402	0.8	0.683
6	0.2919	0.7391	1.031	0.5288	0.29	0.373	0.6211	0.7809
7	0.2413	0.4145	1.0778	0.7836	0.3155	0.2617	1.1276	0.7808
8	0.1696	0.7954	1.0456	0.538	0.1785	0.4238	1.1182	0.726
9	0.2412	0.8813	1.0937	0.4014	0.3293	0.3998	1.9201	0.6492
10	0.2078	0.9006	1.4155	0.3746	0.2704	0.6108	1.7571	0.6505
Average (SD)	0.2594 (0.085)	0.5995 (0.275)	1.0601 (0.186)	0.5904 (0.132)	0.279 (0.089)	0.4058 (0.166)	1.1111 (0.422)	0.7116 (0.051)
Model	M43				M44			
Trait	T1 (GY)		T2 (Testwt)		T1 (GY)		T2 (Testwt)	
PT	MSE	PC	MSE	PC	MSE	PC	MSE	PC
1	0.1717	0.5602	0.9354	0.7015	0.1716	0.5438	0.9623	0.7053
2	0.25	0.4885	0.7258	0.6987	0.2515	0.4383	0.7536	0.6907
3	0.2344	0.5482	1.1496	0.7698	0.2324	0.5182	1.2674	0.7631
4	0.4956	0.0226	1.0861	0.6277	0.4852	−0.055	1.1752	0.5615
5	0.274	0.2154	0.806	0.6778	0.2738	0.1416	0.8357	0.66
6	0.2895	0.3758	0.6211	0.7803	0.2909	0.3468	0.6096	0.7864
7	0.2925	0.319	1.0604	0.7936	0.3001	0.2746	1.0997	0.7856
8	0.1782	0.4247	1.1177	0.7261	0.1828	0.3739	1.1289	0.7139
9	0.324	0.4071	1.9091	0.6488	0.3306	0.3438	1.9212	0.6499
10	0.2974	0.5764	1.8732	0.6506	0.2874	0.5483	1.9158	0.6434
Average (SD)	0.2807 (0.091)	0.3938 (0.173)	1.1284 (0.439)	0.7075 (0.059)	0.2806 (0.088)	0.3474 (0.191)	1.1669 (0.445)	0.696 (0.071)

Model (5.6) with $\boldsymbol{\Sigma}_E = \boldsymbol{I}_I$ is referred to as M4, and if additionally, $\boldsymbol{\Sigma}_{2T} = \text{Diag}\left(\sigma_{1T_1}^2, \sigma_{1T_2}^2\right)$, the model is referred to as M42. Model M42 but with $\boldsymbol{\Sigma}_{1T} = \text{Diag}\left(\sigma_{1T_1}^2, \sigma_{1T_2}^2\right)$ is referred to as M43, and model M44 is the same as M43 but with $\boldsymbol{R} = \text{Diag}\left(\sigma_{e_1}^2, \sigma_{e_2}^2\right)$. The mean squared error of prediction (MSE) and Pearson's correlation (PC) for each trait in each partition are reported. SD is the standard deviation model M4. For the same trait (GY), the PC of model M4 was 47.73%, 52.24%, and 72.57%, greater than the PC of models M42, M43, and M44, respectively.

For the second trait (Testwt), model M4 also showed the best performance, but only under the MSE criteria: the MSE of models M42, M43, and M44 were 4.81%, 6.44%, and 10.08%, respectively, greater than the MSE corresponding to model M4, also suggesting an increasing degradation pattern in the MSE performance as the model became simpler with fewer parameters to estimate in relation to M4. In terms of PC, the best performance was achieved with model M42, which gave 20.54%, 0.583%, and 2.247% greater performances than models M4, M43, and M44, respectively.

Appendix 7 shows the R code used to reproduce the results in Table 5.4 with the sommer package. At the end of this code, we also included the basic code to explore other variance–covariance structures. Specifically, this is the code used to explore the model with heterogeneous genetic variance–covariance matrix, $\boldsymbol{\Sigma}_{2T}$, across environments, that is, $\boldsymbol{g}_{2i} \sim N(\boldsymbol{0}, \boldsymbol{G} \otimes \boldsymbol{\Sigma}_{2iT})$, $i = 1, \ldots, I$, which are assumed independent across environments.

5.6 Final Comments

The multi-trait linear model proposed by Henderson and Quaas (1976) in animal breeding can bring benefits in comparison to single-trait modeling for the improvement of prediction accuracy when incorporating correlated traits, as well as for obtaining an optimal and simplified total merit selection index (Okeke et al. 2017).

When the goal is to predict difficult or expensive traits that are correlated with inexpensive secondary traits, the use of multi-trait models could be helpful in developing better genomic selection strategies. Similarly, improvement of the accuracy of prediction for low-heritability key traits can follow from the use of high-heritability secondary traits (Jia and Jannink 2012; Muranty et al. 2015). Furthermore, this can be combined with the information of traits obtained using the speed breeding methodology to shorten the breeding cycles and accelerate breeding programs (Ghosh et al. 2018; Watson et al. 2019).

While the advantage of the multi-trait model is clearly documented, larger data sets and more computing resources are required, as there are additional parameters that need to be estimated (genetic and error covariances), which may affect the accuracy of genomic prediction. Additionally, convergence problems often arise when implementing complex mixed linear models and especially when small data sets are used.

Although the application of multi-trait models can be easily adapted with regard to genetic correlation, heritability, training population composition, and size of data sets, some other factors need to be carefully considered when using these methods for the improvement of genomic accuracy predictions (Lorenz and Smith 2015; Covarrubias-Pazaran et al. 2018). For example, biased and suboptimal choices between univariate and multi-trait models can result from using auxiliary traits that are measured on individuals to be tested, but appropriate cross-validations strategies could be helpful in determining the usefulness of combining the multi-trait information with multi-trait models (Runcie and Cheng 2019).

Appendix 1

$$
\begin{aligned}
f_{b|Y}(b|y) &\propto f_{Y|b}(y|b)f_b(b)\\
&\propto \frac{1}{(2\pi\sigma^2)^{-\frac{n}{2}}}\exp\left[-\frac{1}{2\sigma^2}(y-X\beta-Zb)^{\mathrm{T}}(y-X\beta-Zb)-\frac{1}{2}b^{\mathrm{T}}D^{-1}b\right]\\
&\propto \exp\left[-\frac{1}{2}b^{\mathrm{T}}(D^{-1}+\sigma^{-2}Z^{\mathrm{T}}Z)b-\sigma^{-2}(y-X\beta)^{\mathrm{T}}Zb\right]\\
&\propto \exp\left[-\frac{1}{2}\left(b-\widetilde{b}\right)^{\mathrm{T}}\widetilde{D}^{-1}\left(b-\widetilde{b}\right)\right]
\end{aligned}
$$

$\widetilde{D}=\left(D^{-1}+\sigma^{-2}Z^{\mathrm{T}}Z\right)^{-1}$ and $\widetilde{b}=\sigma^{-2}\widetilde{D}Z^{\mathrm{T}}(y-X\beta)$. So $b \mid Y=y \sim N_q\left(\widetilde{b},\widetilde{D}\right)$.

Appendix 2

Because $(ABA^{\mathrm{T}}+C)^{-1}=C^{-1}-C^{-1}A(A^{\mathrm{T}}C^{-1}A+B^{-1})^{-1}A^{\mathrm{T}}C^{-1}$ (Johnson and Wichern (2002)), then

$$
\widetilde{D}=\left(D+\sigma^{-2}Z^{\mathrm{T}}Z\right)^{-1}=D-DZ^{\mathrm{T}}\left(ZDZ^{\mathrm{T}}+\sigma^2I_n\right)^{-1}ZD
$$

and thus

$$
\begin{aligned}
\mathrm{tr}\left(\widetilde{D}^{(t)}D^{-1}\right)&=\mathrm{tr}\left[\left(D^{(t)}-D^{(t)}Z^{\mathrm{T}}\left(ZD^{(t)}Z^{\mathrm{T}}+\sigma^{2(t)}I_n\right)^{-1}ZD^{(t)}\right)D^{-1}\right]\\
&=\sum_{k=1}^{K}\frac{\sigma_k^{2(t)}}{\sigma_k^2}\left[q_k-\sigma_k^{2(t)}\mathrm{tr}\left(A_kZ_k^TV^{-(t)}Z_k\right)\right],
\end{aligned}
$$

where $V^{-(t)}=(ZD^{(t)}Z^{\mathrm{T}}+\sigma^{2(t)}I_n)^{-1}$.

Appendix 3

Because

$$\exp\left[-\frac{1}{2}(\boldsymbol{y}-\boldsymbol{X\beta})^{\mathrm{T}}\boldsymbol{V}^{-1}(\boldsymbol{y}-\boldsymbol{X\beta})\right]=\exp\left\{-\frac{1}{2}\left[\boldsymbol{\beta}^{\mathrm{T}}\boldsymbol{X}^{\mathrm{T}}\boldsymbol{V}^{-1}\boldsymbol{X\beta}-2\boldsymbol{y}^{\mathrm{T}}\boldsymbol{V}^{-1}\boldsymbol{X\beta}+\boldsymbol{y}^{\mathrm{T}}\boldsymbol{V}^{-1}\boldsymbol{y}\right]\right\}$$

$$=\exp\left\{-\frac{1}{2}\left(\boldsymbol{\beta}-\widetilde{\boldsymbol{\beta}}\right)^{\mathrm{T}}\boldsymbol{X}^{\mathrm{T}}\boldsymbol{V}^{-1}\boldsymbol{X}\left(\boldsymbol{\beta}-\widetilde{\boldsymbol{\beta}}\right)+\frac{1}{2}\widetilde{\boldsymbol{\beta}}^{\mathrm{T}}\boldsymbol{X}^{\mathrm{T}}\boldsymbol{V}^{-1}\boldsymbol{X}\widetilde{\boldsymbol{\beta}}-\frac{1}{2}\boldsymbol{y}^{\mathrm{T}}\boldsymbol{V}^{-1}\boldsymbol{y}\right\},$$

where $\widetilde{\boldsymbol{\beta}}=\left(\boldsymbol{X}^{\mathrm{T}}\boldsymbol{V}^{-1}\boldsymbol{X}\right)^{-1}\boldsymbol{X}^{\mathrm{T}}\boldsymbol{V}^{-1}\boldsymbol{y}$,

$$\begin{aligned}
f(\boldsymbol{\theta}|\boldsymbol{y}) &\propto \int \frac{|\boldsymbol{V}|^{-\frac{1}{2}}}{(2\pi)^{\frac{n}{2}}}\exp\left[-\frac{1}{2}(\boldsymbol{y}-\boldsymbol{X\beta})^{\mathrm{T}}\boldsymbol{V}^{-1}(\boldsymbol{y}-\boldsymbol{X\beta})\right]d\boldsymbol{\beta}\\
&\propto |\boldsymbol{V}|^{-\frac{1}{2}}\left|\boldsymbol{X}^{\mathrm{T}}\boldsymbol{V}^{-1}\boldsymbol{X}\right|^{-1/2}\exp\left\{-\frac{1}{2}\left[\boldsymbol{y}^{\mathrm{T}}\boldsymbol{V}^{-1}\boldsymbol{y}-\widetilde{\boldsymbol{\beta}}^{\mathrm{T}}\boldsymbol{X}^{\mathrm{T}}\boldsymbol{V}^{-1}\boldsymbol{X}\widetilde{\boldsymbol{\beta}}\right]\right\}\\
&\propto |\boldsymbol{V}|^{-\frac{1}{2}}\left|\boldsymbol{X}^{\mathrm{T}}\boldsymbol{V}^{-1}\boldsymbol{X}\right|^{1/2}\exp\left\{-\frac{1}{2}\left[\boldsymbol{y}^{\mathrm{T}}\boldsymbol{V}^{-1}\boldsymbol{y}-2\widetilde{\boldsymbol{\beta}}^{\mathrm{T}}\boldsymbol{X}^{\mathrm{T}}\boldsymbol{V}^{-1}\boldsymbol{X}\widetilde{\boldsymbol{\beta}}+\widetilde{\boldsymbol{\beta}}^{\mathrm{T}}\boldsymbol{X}^{\mathrm{T}}\boldsymbol{V}^{-1}\boldsymbol{X}\widetilde{\boldsymbol{\beta}}\right]\right\}\\
&\propto |\boldsymbol{V}|^{-\frac{1}{2}}\left|\boldsymbol{X}^{\mathrm{T}}\boldsymbol{V}^{-1}\boldsymbol{X}\right|^{-1/2}\exp\left\{-\frac{1}{2}\left(\boldsymbol{y}-\boldsymbol{X}\widetilde{\boldsymbol{\beta}}\right)^{\mathrm{T}}\boldsymbol{V}^{-1}\left(\boldsymbol{y}-\boldsymbol{X}\widetilde{\boldsymbol{\beta}}\right)\right\}
\end{aligned}$$

Appendix 4

R code for Example 1:

```
rm(list=ls())
library(sommer)
load('dat_ls_E1.RData',verbose=TRUE)
#Phenotypic data
dat_F = dat_ls$dat_F
head(dat_F)
#Marker data
dat_M = dat_ls$dat_M
dim(dat_M)
dat_F = transform(dat_F, GID = as.character(GID))
head(dat_F,5)
#Matrix design of markers
Pos = match(dat_F$GID,row.names(dat_M))
XM = dat_M[Pos,]
XM = scale(XM)
#Genomic relationship matrix derived from markers
dat_M = scale(dat_M)
G = tcrossprod(dat_M)/dim(dat_M)[2]
```

```
dat_F$GID = factor(dat_F$GID,levels=row.names(G))
dat_F = dat_F[order(dat_F$Env,dat_F$GID),]

#10 random partitions
K = 10
n = dim(dat_F)[1]
set.seed(1)
PT = replicate(K,sample(n,0.20*n))

#Example 1
#GBLUP model
Tab = data.frame(PT = 1:K,MSEP = NA)
set.seed(1)
for(k in 1:K)
{
 Pos_tst = PT[,k]
 dat_F$y_NA = dat_F$y
 dat_F$y_NA[Pos_tst] = NA
 #M1
 A = mmer(y_NA ~ 1,na.method.Y='include', random= ~ vs(GID,Gu=G),
         rcov= ~ vs(units), data=dat_F,verbose=FALSE)
 #BLUPs
 #bv = A$U$`u:GID`$y_NA
 yp = fitted(A)$dataWithFitted$y_NA.fitted
 #Prediction of testing
 yp_ts = yp[Pos_tst]
 #MSEP and Cor
 Tab$MSEP[k] = mean((dat_F$y[Pos_tst]-yp_ts)^2)
 Tab$Cor[k] = cor(dat_F$y[Pos_tst],yp_ts)

 #M10
 A2 = mmer(y_NA ~ 1,na.method.Y='include', random= ~ vs(GID),
          rcov= ~ vs(units), data=dat_F,verbose=FALSE)
 yp2 = fitted(A2)$dataWithFitted$y_NA.fitted
 #Prediction of testing
 yp2_ts = yp2[Pos_tst]
 #MSEP
 Tab$MSEP10[k] = mean((dat_F$y[Pos_tst]-yp2_ts)^2)
 Tab$Cor10[k] = cor(dat_F$y[Pos_tst],yp2_ts)

}
```

Appendix 5

```
#Example 2
rm(list=ls())
library(sommer)
load('dat_ls_E1.RData',verbose=TRUE)
```

```
#Phenotypic data
dat_F = dat_ls$dat_F
head(dat_F)
#Marker data
dat_M = dat_ls$dat_M
dim(dat_M)
dat_F = transform(dat_F, GID = as.character(GID))
head(dat_F,5)
#Matrix design of markers
Pos = match(dat_F$GID,row.names(dat_M))
XM = dat_M[Pos,]
XM = scale(XM)
#Genomic relationship matrix derived from markers
dat_M = scale(dat_M)
G = tcrossprod(dat_M)/dim(dat_M)[2]

dat_F$GID = factor(dat_F$GID,levels=row.names(G))
dat_F = dat_F[order(dat_F$Env,dat_F$GID),]

#10 random partitions
K = 10
n = dim(dat_F)[1]
set.seed(1)
PT = replicate(K,sample(n,0.20*n))

#Model (5.4)
#Y = mu + Env + GID + GID:Env
Tab = data.frame(PT = 1:K,MSEP = NA)
set.seed(1)
dat_F$Env_GID = paste(dat_F$Env,dat_F$GID,sep='_')
GE = kronecker(diag(length(unique(dat_F$Env))),G)
rnGWE = expand.grid(row.names(G),unique(dat_F$Env))
row.names(GE) = paste(rnGWE[,2],rnGWE[,1],sep='_')
colnames(GE) = row.names(GE)
for(k in 1:K)
{
 Pos_tst = PT[,k]
 dat_F$y_NA = dat_F$y
 dat_F$y_NA[Pos_tst] = NA
 #M2
 A = mmer(y_NA ~ Env,na.method.Y='include',
      random= ~ vs(GID,Gu=G) + vs(Env_GID,Gu = GE),
      rcov= ~ vs(units),
      data=dat_F,verbose=FALSE)
 yp = fitted(A)$dataWithFitted$y_NA.fitted
 #Prediction of testing
 yp_ts = yp[Pos_tst]
 #MSEP
 Tab$MSEP[k] = mean((dat_F$y[Pos_tst]-yp_ts)^2)
 Tab$Cor[k] = cor(dat_F$y[Pos_tst],yp_ts)

 #M20
 A20 = mmer(y_NA ~ Env,na.method.Y='include',
```

```
      random= ~ vs(GID)+vs(Env_GID),
      rcov= ~ vs(units),
      data=dat_F,verbose=FALSE)
 yp20 = fitted(A20)$dataWithFitted$y_NA.fitted
 #Prediction of testing
 yp20_ts = yp20[Pos_tst]
 Tab$MSEP20[k] = mean((dat_F$y[Pos_tst]-yp20_ts)^2)
 Tab$Cor20[k] = cor(dat_F$y[Pos_tst],yp20_ts)

 #M11
 A = mmer(y_NA ~ Env,na.method.Y='include',
        random= ~ vs(GID,Gu=G),
        rcov= ~ vs(units),
        data=dat_F,verbose=FALSE)
 yp = fitted(A)$dataWithFitted$y_NA.fitted
 #Prediction of testing
 yp_ts = yp[Pos_tst]
 Tab$MSEP11[k] = mean((dat_F$y[Pos_tst]-yp_ts)^2)
 Tab$Cor11[k] = cor(dat_F$y[Pos_tst],yp_ts)
 #M10a: Model1 (5.4) plus environment effects but with G = I_J
 A = mmer(y_NA ~ Env,na.method.Y='include',
       random= ~ vs(GID),
       rcov= ~ vs(units),
       data=dat_F,verbose=FALSE)
 yp = fitted(A)$dataWithFitted$y_NA.fitted
 #Prediction of testing
 yp_ts = yp[Pos_tst]
 Tab$MSEP10a[k] = mean((dat_F$y[Pos_tst]-yp_ts)^2)
 Tab$Cor10a[k] = cor(dat_F$y[Pos_tst],yp_ts)

}

#Basic code to implement model (5.4) with an unstructured form for
Sigma_E
A = mmer(y~ Env, na.method.Y='include',
          random= ~ vs(GID,Gu=G)+vs(us(Env),GID,Gu=G),
          rcov= ~ vs(units), data=dat_F)
#Basic code to implement model (5.4) with heterogeneous environment
variances
A = mmer(y ~ Env, ,
          random= ~ vs(GID,Gu=G) +
                    vs(ds(Env),GID,Gu=G),#or vs(at(Env),GID,Gu=G)
          rcov= ~ vs(units), data=dat_F)
#Y = mu + Env + GID + GID:ENV
#Y = mu + Env + GID + GID:ENV
#Basic code to implement model (5.4) with heterogeneous environment and
residuals variances
A = mmer(y ~ Env,
      random= ~ vs(GID,Gu=G) + vs(ds(Env),GID,Gu=G),
      rcov= ~ vs(ds(Env),units),
      data=dat_F)
```

Appendix 6

```
rm(list=ls())
library(sommer)
load('dat_ls_E1.RData',verbose=TRUE)
#Phenotypic data
dat_F = dat_ls$dat_F
head(dat_F)
#Marker data
dat_M = dat_ls$dat_M
dim(dat_M)
dat_F = transform(dat_F, GID = as.character(GID))
head(dat_F,5)
#Matrix design of markers
Pos = match(dat_F$GID,row.names(dat_M))
XM = dat_M[Pos,]
XM = scale(XM)
dim(XM)
n = dim(dat_F)[1]
#Genomic relationship matrix derived from markers
dat_M = scale(dat_M)
G = tcrossprod(dat_M)/dim(dat_M)[2]

dat_F$GID = factor(dat_F$GID,levels=row.names(G))
dat_F = dat_F[order(dat_F$Env,dat_F$GID),]

#10 random partitions
K = 10
set.seed(1)
PT = replicate(K,sample(n,0.20*n))


#Example 3
Tab = data.frame(PT = 1:K)
set.seed(1)
for(k in 1:K)
{
 #M3
 Pos_tst = PT[,k]
 dat_F$y_NA = dat_F$y
 dat_F$y_NA[Pos_tst] = NA
 dat_F$y2_NA = dat_F$y2
 dat_F$y2_NA[Pos_tst] = NA
 A = mmer(cbind(y_NA,y2_NA) ~ 1,na.method.Y='include',
       random= ~ vs(GID,Gu=G) ,
       rcov= ~ vs(units),
       data=dat_F,verbose=FALSE)
 #BLUPs
 b_ls = A$U$`u:GID`
 b_T1 = b_ls$y_NA# Trait 1
 b_T2 = b_ls$y2_NA# Trait 2
```

```
b_mat = cbind(b_T1,b_T2)
Pos = match(dat_F$GID,names(b_ls$y_NA))
#Y "fitted"
yp = A$fitted + b_mat[Pos,]
#Prediction of testing for both traits
yp_ts = yp[Pos_tst,1]# Trait 1
y2p_ts = yp[Pos_tst,2]# Trait 2
#MSEP and Cor
#Trait 1
Tab$MSEP_T1[k] = mean((dat_F$y[Pos_tst]-yp_ts)^2)
Tab$Cor_T1[k] = cor(dat_F$y[Pos_tst],yp_ts)
#Trait 2
Tab$MSEP_T2[k] = mean((dat_F$y2[Pos_tst]-y2p_ts)^2)
Tab$Cor_T2[k] = cor(dat_F$y2[Pos_tst],y2p_ts)
#M32: Sigma_T diagonal
A = mmer(cbind(y_NA,y2_NA) ~ 1,na.method.Y='include',
     random= ~ vs(GID,Gu=G,Gtc=diag(2)),
     rcov= ~ vs(units),
     data=dat_F,verbose=FALSE)
#BLUPs
b_ls = A$U$`u:GID`
b_T1 = b_ls$y_NA# Trait 1
b_T2 = b_ls$y2_NA# Trait 2
b_mat = cbind(b_T1,b_T2)
Pos = match(dat_F$GID,names(b_ls$y_NA))
#Y "fitted"
yp = A$fitted + b_mat[Pos,]
#Prediction of testing
yp_ts = yp[Pos_tst,1]# Trait 1
y2p_ts = yp[Pos_tst,2]# Trait 2
#MSEP and Cor
#Trait 1
Tab$MSEP_T1_32[k] = mean((dat_F$y[Pos_tst]-yp_ts)^2)
Tab$Cor_T1_32[k] = cor(dat_F$y[Pos_tst],yp_ts)
#Trait 2
Tab$MSEP_T2_32[k] = mean((dat_F$y2[Pos_tst]-y2p_ts)^2)
Tab$Cor_T2_32[k] = cor(dat_F$y2[Pos_tst],y2p_ts)


# #M3
# #dat_F$y2_NA = dat_F$y2
# dat_F$y_NA = dat_F$y# Trait T1 is not NA
# A = mmer(cbind(y_NA,y2_NA) ~ 1,na.method.Y='include',
#       random= ~ vs(GID,Gu=G),
#       rcov= ~ vs(units),#tolparinv = 1e-2,
#       data=dat_F,verbose=FALSE)
# #BLUPs
# b_ls = A$U$`u:GID`
# b_T1 = b_ls$y_NA# Trait 1
# b_T2 = b_ls$y2_NA# Trait 2
# b_mat = cbind(b_T1,b_T2)
# Pos = match(dat_F$GID,names(b_ls$y_NA))
# #Y "fitted"
# yp = A$fitted + b_mat[Pos,]
```

```
 # #Prediction of testing
 # yp_ts = yp[Pos_tst,1]# Trait 1
 # y2p_ts = yp[Pos_tst,2]# Trait 2
 # #MSEP
 # Tab$MSEP_T1_31[k] = mean((dat_F$y[Pos_tst]-yp_ts)^2)
 # Tab$Cor_T1_31[k] = cor(dat_F$y[Pos_tst],yp_ts)
 # Tab$MSEP_T2_31[k] = mean((dat_F$y2[Pos_tst]-y2p_ts)^2)
 # Tab$Cor_T2_31[k] = cor(dat_F$y2[Pos_tst],y2p_ts)
}
```

Appendix 7

```
#Example 4
rm(list=ls())
library(sommer)
load('dat_ls_E4.RData',verbose=TRUE)
#Phenotypic data
dat_F = dat_ls$dat_F
dat_F = transform(dat_F, GID = as.character(GID))
#Genomic relationship matrix derived from markers
G = dat_ls$G
dat_F$GID = factor(dat_F$GID,levels=row.names(G))
dat_F = dat_F[order(dat_F$Env,dat_F$GID),]
#Fitting model M4 with data set
#A = mmer(cbind(GY,TESTWT) ~ Env,na.method.Y='include',
#      random= ~ vs(GID,Gu=G) + vs(Env_GID,Gu=GE) ,
#      rcov= ~ vs(units),
#      data=dat_F,verbose=FALSE)
#Example 4
#10 random partitions
n = dim(dat_F)[1] ;K = 10
set.seed(1)
PT = replicate(K,sample(n,0.20*n))
#Model 5.6
#Y = mu + Env + GID + GID:Env + e
Tab = data.frame(PT = 1:K)
set.seed(1)
dat_F$Env_GID = paste(dat_F$Env,dat_F$GID,sep='_')
GE = kronecker(diag(length(unique(dat_F$Env))),G)
rnGWE = expand.grid(row.names(G),unique(dat_F$Env))
row.names(GE) = paste(rnGWE[,2],rnGWE[,1],sep='_')
colnames(GE) = row.names(GE)
for(k in 1:K)
{
 Pos_tst = PT[,k]
 #M4
 Pos_tst = PT[,k]
 dat_F$GY_NA = dat_F$GY
 dat_F$GY_NA[Pos_tst] = NA
```

```
dat_F$TESTWT_NA = dat_F$TESTWT
dat_F$TESTWT_NA[Pos_tst] = NA
A = mmer(cbind(GY_NA,TESTWT_NA ) ~ Env,na.method.Y='include',
      random= ~ vs(GID,Gu=G) + vs(Env_GID,Gu=GE) ,
      rcov= ~ vs(units),tolparinv = 1,
      data=dat_F,verbose=FALSE)
#BLUPs
b_ls = A$U$`u:GID`
b_T1 = b_ls$GY_NA# Trait 1
b_T2 = b_ls$TESTWT_NA# Trait 2
b_mat = cbind(b_T1,b_T2)
Pos = match(dat_F$GID,names(b_ls$TESTWT))
#Y "fitted"
yp = data.frame(A$fitted + b_mat[Pos,] )
colnames(yp) = names(b_ls)
plot(dat_F$GY,yp$GY_NA);abline(a=0,b=1)
#Prediction of testing of both traits
yp_tst = yp[Pos_tst,]
#plot(dat_F$y2,yp[,2]); abline(a=0,b=1)
#MSEP and Cor
#Trait 1
Tab$MSEPT1[k] = mean((dat_F$GY[Pos_tst]-yp_tst$GY_NA)^2)
Tab$CorT1[k] = cor(dat_F$GY[Pos_tst],yp_tst$TESTWT_NA)
#Trait 2
Tab$MSEPT2[k] = mean((dat_F$TESTWT[Pos_tst]-yp_tst$TESTWT_NA)^2)
Tab$CorT2[k] = cor(dat_F$TESTWT[Pos_tst],yp_tst$TESTWT_NA)

#M42
A42 = mmer(cbind(GY_NA,TESTWT_NA) ~ Env,na.method.Y='include',
      random= ~ vs(GID,Gu=G) + vs(Env_GID,Gu=GE,Gtc=diag(2)) ,
      rcov= ~ vs(units),
      data=dat_F,verbose=FALSE)
#BLUPs
b_ls = A42$U$`u:GID`
b_T1 = b_ls$GY_NA# Trait 1
b_T2 = b_ls$TESTWT_NA# Trait 2
b_mat = cbind(b_T1,b_T2)
Pos = match(dat_F$GID,names(b_ls$TESTWT))
#Y "fitted"
yp = data.frame(A$fitted + b_mat[Pos,] )
colnames(yp) = names(b_ls)
#plot(dat_F$GY,yp$GY_NA);abline(a=0,b=1)
#Prediction of testing of both traits
yp_tst = yp[Pos_tst,]
#MSEP and Cor
#Trait 1
Tab$MSEPT1_42[k] = mean((dat_F$GY[Pos_tst]-yp_tst$GY_NA)^2)
Tab$CorT1_42[k] = cor(dat_F$GY[Pos_tst],yp_tst$TESTWT_NA)
#Trait 2
Tab$MSEPT2_42[k] = mean((dat_F$TESTWT[Pos_tst]-yp_tst$TESTWT_NA)
^2)
Tab$CorT2_42[k] = cor(dat_F$TESTWT[Pos_tst],yp_tst$TESTWT_NA)
```

```
#M43
A43 = mmer(cbind(GY_NA,TESTWT_NA) ~ Env,na.method.Y='include',
      random= ~ vs(GID,Gu=G,Gtc=diag(2)) +
           vs(Env_GID,Gu=GE,Gtc=diag(2)) ,
      rcov= ~ vs(units),
      data=dat_F,verbose=FALSE)
#BLUPs
b_ls = A43$U$`u:GID`
b_T1 = b_ls$GY_NA# Trait 1
b_T2 = b_ls$TESTWT_NA# Trait 2
b_mat = cbind(b_T1,b_T2)
Pos = match(dat_F$GID,names(b_ls$TESTWT))
#Y "fitted"
yp = data.frame(A$fitted + b_mat[Pos,] )
colnames(yp) = names(b_ls)
#plot(dat_F$GY,yp$GY_NA);abline(a=0,b=1)
#Prediction of testing of both traits
yp_tst = yp[Pos_tst,]
#MSEP and Cor
#Trait 1
Tab$MSEPT1_43[k] = mean((dat_F$GY[Pos_tst]-yp_tst$GY_NA)^2)
Tab$CorT1_43[k] = cor(dat_F$GY[Pos_tst],yp_tst$TESTWT_NA)
#Trait 2
Tab$MSEPT2_43[k] = mean((dat_F$TESTWT[Pos_tst]-yp_tst$TESTWT_NA)
^2)
Tab$CorT2_43[k] = cor(dat_F$TESTWT[Pos_tst],yp_tst$TESTWT_NA)
 A44 = mmer(cbind(GY_NA,TESTWT_NA) ~ Env,na.method.Y='include',
      random= ~ vs(GID,Gu=G,Gtc=diag(2)) +
       vs(Env_GID,Gu=GE,Gtc=diag(2)) ,
      rcov= ~ vs(units,Gtc=diag(2)),
      data=dat_F,verbose=FALSE)
#A44$sigma
#BLUPs
b_ls = A44$U$`u:GID`
b_T1 = b_ls$GY_NA# Trait 1
b_T2 = b_ls$TESTWT_NA# Trait 2
b_mat = cbind(b_T1,b_T2)
Pos = match(dat_F$GID,names(b_ls$TESTWT))
#Y "fitted"
yp = data.frame(A$fitted + b_mat[Pos,] )
colnames(yp) = names(b_ls)
#plot(dat_F$GY,yp$GY_NA);abline(a=0,b=1)
#Prediction of testing of both traits
yp_tst = yp[Pos_tst,]
#MSEP and Cor
#Trait 1
Tab$MSEPT1_44[k] = mean((dat_F$GY[Pos_tst]-yp_tst$GY_NA)^2)
Tab$CorT1_44[k] = cor(dat_F$GY[Pos_tst],yp_tst$TESTWT_NA)
#Trait 2
Tab$MSEPT2_44[k] = mean((dat_F$TESTWT[Pos_tst]-yp_tst$TESTWT_NA)
^2)
Tab$CorT2_44[k] = cor(dat_F$TESTWT[Pos_tst],yp_tst$TESTWT_NA)
cat('k=',k,'\n')

}
```

```
#Model 5.6 with Sigma_2T different for each Env
A4 = mmer(cbind(GY,TESTWT)~Env,
      random = ~vs(GID,Gu=G)+vs(ds(Env),GID,Gu=G),data=dat_F,
      rcov= ~ vs(units))
```

References

Araus JL, Cairns JE (2014) Field high-throughput phenotyping: the new crop breeding frontier. Trends Plant Sci 19(1):52–61

Bates D, Maechler M, Bolker B, Walker S (2015) Fitting linear mixed-effects models using lme4. J Stat Softw 67(1):1–48

Borman S (2004) The expectation maximization algorithm: a short tutorial. https://www.lri.fr/~sebag/COURS/EM_algorithm.pdf

Brown H, Prescott R (2014) Applied mixed models in medicine. John Wiley & Sons, Hoboken, NJ

Budhlakoti N, Mishra DC, Rai A, Lal SB, Chaturvedi KK, Kumar RR (2019) A comparative study of single-trait and multi-trait genomic selection. J Comput Biol 26(10):1100–1112

Burgueño J, de los Campos G, Weigel K, Crossa J (2012) Genomic prediction of breeding values when modeling genotype × environment interaction using pedigree and dense molecular markers. Crop Sci 52(2):707–719

Cabrera-Bosquet L, Crossa J, von Zitzewitz J, Serret MD, Luis Araus J (2012) High-throughput phenotyping and genomic selection: the frontiers of crop breeding converge F. J Integr Plant Biol 54(5):312–320

Calus MP, Veerkamp RF (2011) Accuracy of multi-trait genomic selection using different methods. Genet Select Evol 43(1):26. https://doi.org/10.1186/1297-9686-43-26

Cappa EP, de Lima BM, da Silva-Junior OB, Garcia CC, Mansfield SD, Grattapaglia D (2019) Improving genomic prediction of growth and wood traits in Eucalyptus using phenotypes from non-genotyped trees by single-step GBLUP. Plant Sci 284:9–15

Covarrubias-Pazaran G (2016) Genome-assisted prediction of quantitative traits using the R package sommer. PLoS One 11(6):e0156744

Covarrubias-Pazaran G (2018) Software update: moving the R package sommer to multivariate mixed models for genome-assisted prediction. https://doi.org/10.1101/354639

Covarrubias-Pazaran G, Schlautman B, Diaz-Garcia L, Grygleski E, Polashock J, Johnson-Cicalese J et al (2018) Multivariate GBLUP improves accuracy of genomic selection for yield and fruit weight in biparental populations of Vaccinium macrocarpon Ait. Front Plant Sci 9:1310

Crossa J, Pérez-Rodríguez P, Cuevas J, Montesinos-López O, Jarquín D, de los Campos G et al (2017) Genomic selection in plant breeding: methods, models, and perspectives. Trends Plant Sci 22(11):961–975

Finch WH, Bolin JE, Kelley K (2019) Multilevel modeling using R. CRC Press, Boca Raton, FL

Ghosh S, Watson A, Gonzalez-Navarro OE, Ramirez-Gonzalez RH, Yanes L, Mendoza-Suárez M et al (2018) Speed breeding in growth chambers and glasshouses for crop breeding and model plant research. Nat Protoc 13(12):2944–2963

Goldstein H (2011) Multilevel statistical models. Wiley, Hoboken, NJ

Harville DA (1974) Bayesian inference for variance components using only error contrasts. Biometrika 61(2):383–385

Harville DA (1977) Maximum likelihood approaches to variance component estimation and to related problems. J Am Stat Assoc 72(358):320–338

Harville DA (1985) Decomposition of prediction error. J Am Stat Assoc 80(389):132–138

Henderson CR (1950) Estimation of genetic parameters. Ann Math Stat 21:309–310

Henderson CR (1975) Best linear unbiased estimation and prediction under a selection model. Biometrics 31:423–447

Henderson CR, Quaas RL (1976) Multiple trait evaluation using relatives' records. J Anim Sci 43 (6):1188–1197

Jarquín D, Crossa J, Lacaze X, Du Cheyron P, Daucourt J, Lorgeou J et al (2014) A reaction norm model for genomic selection using high-dimensional genomic and environmental data. Theor Appl Genet 127(3):595–607

Jennrich RI, Sampson PF (1976) Newton-Raphson and related algorithms for maximum likelihood variance component estimation. Technometrics 18(1):11–17

Jennrich RI, Schluchter MD (1986) Unbalanced repeated-measures models with structured covariance matrices. Biometrics 42:805–820

Jia Y, Jannink JL (2012) Multiple-trait genomic selection methods increase genetic value prediction accuracy. Genetics 192(4):1513–1522

Jiang J, Zhang Q, Ma L, Li J, Wang Z, Liu JF (2015) Joint prediction of multiple quantitative traits using a Bayesian multivariate antedependence model. Heredity 115(1):29–36

Johnson RA, Wichern DW (2002) Applied multivariate statistical analysis. Prentice Hall, Upper Saddle River, NJ

Krause MR, González-Pérez L, Crossa J, Pérez-Rodríguez P, Montesinos-López O, Singh RP et al (2019) Hyperspectral reflectance-derived relationship matrices for genomic prediction of grain yield in wheat. G3 9(4):1231–1247

Laird NM, Ware JH (1982) Random-effects models for longitudinal data. Biometrics 38:963–974

Leyland AH, Goldstein H (2001) Multilevel modelling of health statistics. Wiley, Hoboken, NJ

Lindstrom MJ, Bates DM (1988) Newton–Raphson and EM algorithms for linear mixed-effects models for repeated-measures data. J Am Stat Assoc 83(404):1014–1022

Lorenz AJ, Smith KP (2015) Adding genetically distant individuals to training populations reduces genomic prediction accuracy in barley. Crop Sci 55(6):2657–2667

Martini JW, Crossa J, Toledo FH, Cuevas J (2020) On Hadamard and Kronecker products in covariance structures for genotype × environment interaction. Plant Genome 13:e20033

McLean RA, Sanders WL, Stroup WW (1991) A unified approach to mixed linear models. Am Stat 45(1):54–64

Meeker W, Hong Y, Escobar L (2011) Degradation models and analyses. In: Encyclopedia of statistical sciences. Wiley, Hoboken, NJ. https://doi.org/10.1002/0471667196.ess7148

Meuwissen THE, Hayes BJ, Goddard ME (2001) Prediction of total genetic values using genome-wide dense marker maps. Genetics 157:1819–1829

Montesinos-López OA, Montesinos-López A, Crossa J, Toledo FH, Pérez-Hernández O, Eskridge KM, Rutkoski J (2016) A genomic Bayesian multi-trait and multi-environment model. G3 6 (9):2725–2744

Muranty H, Troggio M, Sadok IB, Al Rifaï M, Auwerkerken A, Banchi E et al (2015) Accuracy and responses of genomic selection on key traits in apple breeding. Horticult Res 2(1):1–12

Okeke UG, Akdemir D, Rabbi I, Kulakow P, Jannink JL (2017) Accuracies of univariate and multivariate genomic prediction models in African cassava. Genet Sel Evol 49(1):88

Patterson HD, Thompson R (1971) Recovery of inter-block information when block sizes are unequal. Biometrika 58:545–554

Piepho HP, Möhring J, Melchinger AE, Büchse A (2008) BLUP for phenotypic selection in plant breeding and variety testing. Euphytica 161(1–2):209–228

Pinheiro JC, Bates DM (2000) Mixed-effects models in S and S-PLUS. Springer, New York

Poland J, Endelman J, Dawson J, Rutkoski J, Wu S, Manes Y et al (2012) Genomic selection in wheat breeding using genotyping-by-sequencing. Plant Genome 5(3):103–113

Raudenbush SW, Bryk AS (2002) Hierarchical linear models: applications and data analysis methods. Sage Publications, Inc, Thousand Oaks, CA

Rencher AC (2008) Linear models in statistics. Wiley, Hoboken, NJ

Robinson GK (1991) That BLUP is a good thing: the estimation of random effects. Stat Sci 6 (1):15–32

Runcie D, Cheng H (2019) Pitfalls and remedies for cross validation with multi-trait genomic prediction methods. G3 9(11):3727–3741

Searle SR (1993) Applying the EM algorithm to calculating ML and REML estimates of variance components. In: Paper invited for the 1993 American Statistical Association Meeting, San Francisco

Searle SR, Casella G, McCulloch CE (2006) Variance components. Wiley, Hoboken, NJ

Speelman D, Heylen K, Geeraerts D (eds) (2018) Mixed-effects regression models in linguistics. Springer, New York

Stroup WW (2012) Generalized linear mixed models: modern concepts, methods and applications. CRC Press, Boca Raton, FL

VanRaden PM (2008) Efficient methods to compute genomic predictions. J Dairy Sci 91:4414–4423

Wang X, Xu Y, Hu Z, Xu C (2018) Genomic selection methods for crop improvement: current status and prospects. Crop J 6(4):330–340

Watson A, Hickey LT, Christopher J, Rutkoski J, Poland J, Hayes BJ (2019) Multivariate genomic selection and potential of rapid indirect selection with speed breeding in spring wheat. Crop Sci 59(5):1945–1959

West BT, Welch KB, Galecki AT (2014) Linear mixed models: a practical guide using statistical software. CRC Press, Boca Raton, FL

Zuur A, Ieno EN, Walker N, Saveliev AA, Smith GM (2009) Mixed effects models and extensions in ecology with R. Springer Science & Business Media, New York

6

Genomic-Enabled Prediction: Various Methods

6.1 Bayes Theorem and Bayesian Linear Regression

Unlike classic statistical inference, which assumes that the parameter $\boldsymbol{\theta}$ that defines the model is a fixed unknown quantity, in Bayesian inference it is considered as a random variable whose variation tries to represent the knowledge or ignorance about this, before the data points are collected (Box and Tiao 1992). The probability density function which describes such variation is known as the prior distribution and is an additional component in the specification of the complete model in a Bayesian framework.

Given a data set $\boldsymbol{y}_n = (y_1, \ldots, y_n)$ whose distribution is assumed to be $f(\boldsymbol{y} | \boldsymbol{\theta})$, and a prior distribution for the parameter $\boldsymbol{\theta}$, $f(\boldsymbol{\theta})$, the Bayesian analysis uses the Bayes theorem to combine these two pieces of information to obtain the posterior distribution of the parameters, on which the inference is fully based (Christensen et al. 2011):

$$f(\boldsymbol{\theta}|\boldsymbol{y}) = \frac{f(\boldsymbol{y}, \boldsymbol{\theta})}{f(\boldsymbol{y})} = \frac{f(\boldsymbol{\theta})f(\boldsymbol{y}|\boldsymbol{\theta})}{f(\boldsymbol{y})} \propto f(\boldsymbol{\theta})L(\boldsymbol{\theta}; \boldsymbol{y}),$$

where $f(\boldsymbol{y}) = \int f(\boldsymbol{y} | \boldsymbol{\theta})f(\boldsymbol{\theta})d\boldsymbol{\theta} = E_{\boldsymbol{\theta}}[f(\boldsymbol{y} | \boldsymbol{\theta})]$ is the marginal distribution of $\boldsymbol{\theta}$. This conditional distribution describes what is known about $\boldsymbol{\theta}$ after data is collected and can be thought of as the updated prior knowledge about $\boldsymbol{\theta}$ with the information contained in the data, which is done through the likelihood function $L(\boldsymbol{\theta}; \boldsymbol{y})$ (Box and Tiao 1992).

In general, because the posterior distribution doesn't always have a recognizable form and it is often not easy to simulate from this, numerical approximation methods are employed. Once a sample of the posterior distribution is obtained, estimation of a parameter is often found by averaging the sample values or averaging a function of the sample values when another quantity is of interest. For example, in genomic prediction with dense molecular markers, the main interest is to predict the trait of interest of the non-phenotyped individuals that have only genotypic information, environment variables, or other information (covariates). In this situation, a convenient practice is to include the individuals to be predicted (y_p) in the posterior distribution to be sampled.

Specifically, a standard Bayesian framework for a normal linear regression model (see Chap. 3)

$$Y=\beta_0+\sum_{j=1}^{p}X_j\beta_j+\epsilon \tag{6.1}$$

with ϵ a random error with normal distribution with mean 0 and variance σ^2, is fully specified by assuming the next non-informative prior distribution: $\boldsymbol{\beta}$ and log(σ) approximately independent and locally uniform.

$$f\left(\boldsymbol{\beta},\sigma^2\right)\propto\sigma^{-2} \tag{6.2}$$

which is not a proper distribution because it does not integrate to 1 (Box and Tiao 1992; Gelman et al. 2013). However, when $\boldsymbol{X}$ is of full column rank, the posterior distribution is a proper distribution and is given by

$$\begin{aligned}
f(\boldsymbol{\beta},\sigma^2|\boldsymbol{y},\boldsymbol{X}) &\propto (\sigma^2)^{-\frac{n}{2}}\exp\left[-\frac{1}{2\sigma^2}(\boldsymbol{y}-\boldsymbol{X\beta})^{\mathrm{T}}(\boldsymbol{y}-\boldsymbol{X\beta})\right](\sigma^2)^{-1}\\
&\propto (\sigma^2)^{-\frac{n}{2}}\exp\left[-\frac{1}{2\sigma^2}(\boldsymbol{\beta}^{\mathrm{T}}\boldsymbol{X}^{\mathrm{T}}\boldsymbol{X\beta}-2\boldsymbol{y}^{\mathrm{T}}\boldsymbol{X\beta}+\boldsymbol{y}^{\mathrm{T}}\boldsymbol{y})\right](\sigma^2)^{-1}\\
&\propto (\sigma^2)^{-\frac{p+1}{2}}\exp\left[-\frac{1}{2\sigma^2}\left(\boldsymbol{\beta}-\widetilde{\boldsymbol{\beta}}\right)^{\mathrm{T}}\boldsymbol{X}^{\mathrm{T}}\boldsymbol{X}\left(\boldsymbol{\beta}-\widetilde{\boldsymbol{\beta}}\right)-\frac{1}{2\sigma^2}\left(\boldsymbol{y}^{\mathrm{T}}\boldsymbol{y}-\widetilde{\boldsymbol{\beta}}^{\mathrm{T}}\boldsymbol{X}^{\mathrm{T}}\boldsymbol{X}\widetilde{\boldsymbol{\beta}}\right)\right](\sigma^2)^{-1-(n-p-1)/2}\\
&\propto (\sigma^2)^{-\frac{p+1}{2}}\exp\left[-\frac{1}{2\sigma^2}\left(\boldsymbol{\beta}-\widetilde{\boldsymbol{\beta}}\right)^{\mathrm{T}}\boldsymbol{X}^{\mathrm{T}}\boldsymbol{X}\left(\boldsymbol{\beta}-\widetilde{\boldsymbol{\beta}}\right)\right](\sigma^2)^{-1-(n-p-1)/2}\exp\left[-\frac{1}{2\sigma^2}\boldsymbol{y}^{\mathrm{T}}(\boldsymbol{I}-\boldsymbol{H})\boldsymbol{y}\right]\\
&\propto (\sigma^2)^{-\frac{p+1}{2}}\exp\left[-\frac{1}{2\sigma^2}\left(\boldsymbol{\beta}-\widetilde{\boldsymbol{\beta}}\right)^{\mathrm{T}}\boldsymbol{X}^{\mathrm{T}}\boldsymbol{X}\left(\boldsymbol{\beta}-\widetilde{\boldsymbol{\beta}}\right)\right](\sigma^2)^{-1-\frac{n-p-1}{2}}\exp\left(-\frac{(n-p-1)\widetilde{\sigma}^2}{2\sigma^2}\right),
\end{aligned}$$

where $\widetilde{\boldsymbol{\beta}}=\left(\boldsymbol{X}^{\mathrm{T}}\boldsymbol{X}\right)^{-1}\boldsymbol{X}^{\mathrm{T}}\boldsymbol{y}$, $\widetilde{\sigma}^2=\frac{1}{n-p-1}\boldsymbol{y}^{\mathrm{T}}(\boldsymbol{I}-\boldsymbol{H})\boldsymbol{y}$, and $\boldsymbol{H}=\boldsymbol{X}(\boldsymbol{X}^{\mathrm{T}}\boldsymbol{X})^{-1}\boldsymbol{X}^{\mathrm{T}}$. From here the marginal posterior distribution of σ^2 is $\sigma^2\mid\boldsymbol{y},\boldsymbol{X}\sim IG\left((n-p-1)/2,\frac{(n-p-1)\widetilde{\sigma}^2}{2}\right)$ with mean $\frac{(n-p-1)\widetilde{\sigma}^2}{2}/[(n-p-1)/2]=\widetilde{\sigma}^2$, and given σ^2, the posterior conditional distribution of $\boldsymbol{\beta}$ is given by $\boldsymbol{\beta}\mid\sigma^2,\boldsymbol{y},\boldsymbol{X}\sim N\left(\widetilde{\boldsymbol{\beta}},\sigma^2\left(\boldsymbol{X}^{\mathrm{T}}\boldsymbol{X}\right)^{-1}\right)$.

6.2 Bayesian Genome-Based Ridge Regression

When $p>n$, $\boldsymbol{X}$ is not of full column rank and the posterior of model (6.1) may not be proper (Gelman et al. 2013), so a solution is instead to consider independently proper prior distributions, $\boldsymbol{\beta}\sim N(0,\boldsymbol{I}\sigma^2)$ and $\sigma^2\sim IG(\alpha_0,\alpha_0)$, which for large values of σ^2

(10^6) and small values of α_0 (10^{-3}) is an approximation to the standard non-informative prior given in (6.1) (Christensen et al. 2011). A similar prior specification is taken in genomic prediction where different models are obtained by adopting different prior distributions of the parameters. For example, the Bayesian Linear Ridge Regression (Pérez and de los Campos 2014) with standardized covariates ($X_j's$) is given by

$$Y = \mu + \sum_{j=1}^{p} X_j\beta_j + \epsilon \tag{6.3}$$

with a flat prior for mean parameter (μ), $f(\mu) \propto 1$, which can be approximately specified by $\mu \sim N(0, \sigma_0^2)$, with a large value of σ_0^2 (10^{10}), a multivariate normal distribution with mean vector $\mathbf{0}$ and covariance matrix $\sigma_\beta^2\boldsymbol{I}_p$ on the beta coefficients, $\boldsymbol{\beta}_0 = (\beta_1, \ldots, \beta_p)^{\mathrm{T}} \mid \sigma_\beta^2 \sim N_p\left(\mathbf{0}, \boldsymbol{I}_p\sigma_\beta^2\right)$, and scaled inverse Chi-square distributions as priors for the variance component: $\sigma_\beta^2 \sim \chi^{-2}_{v_\beta,\ S_\beta}$ (prior for the variance of the regression coefficients β_j) and $\sigma^2 \sim \chi^{-2}_{v,S}$ (prior for the variance of random errors, ϵ), where $\chi^{-2}_{v,S}$ denotes the scaled inverse Chi-squared distribution with shape parameter v and scale parameter S. The joint posterior distribution of the parameters in this model, $\boldsymbol{\theta} = \left(\mu, \boldsymbol{\beta}_0^{\mathrm{T}}, \sigma^2, \sigma_\beta^2\right)^{\mathrm{T}}$, is given by

$$\begin{aligned}
f\left(\mu, \boldsymbol{\beta}_0^{\mathrm{T}}, \sigma_\beta^2, \sigma^2 | \boldsymbol{y}, \boldsymbol{X}\right) &\propto L(\mu, \boldsymbol{\beta}_0^{\mathrm{T}}, \sigma^2; \boldsymbol{y}) f(\boldsymbol{\theta}) \\
&\propto L(\mu, \boldsymbol{\beta}_0^{\mathrm{T}}, \sigma^2; \boldsymbol{y}) f(\mu) f\left(\boldsymbol{\beta}_0 | \sigma_\beta^2\right) f\left(\sigma_\beta^2\right) f(\sigma^2) \\
&\propto \frac{1}{(2\pi\sigma^2)^{\frac{n}{2}}} \exp\left[-\frac{1}{2\sigma^2}\|\boldsymbol{y} - \boldsymbol{1}_n\mu - \boldsymbol{X}_1\boldsymbol{\beta}_0\|^2\right] \times \exp\left(-\frac{1}{2\sigma_0^2}\mu^2\right) \\
&\quad \times \frac{1}{\left(\sigma_\beta^2\right)^{\frac{p}{2}}} \exp\left(\left[-\frac{1}{2\sigma_\beta^2}\boldsymbol{\beta}_0^{\mathrm{T}}\boldsymbol{\beta}_0\right]\right) \times \frac{\left(\frac{S_\beta}{2}\right)^{\frac{v_\beta}{2}}}{\Gamma\left(\frac{v_\beta}{2}\right)\left(\sigma_\beta^2\right)^{1+\frac{v_\beta}{2}}} \exp\left(-\frac{S_\beta}{2\sigma_\beta^2}\right) \\
&\quad \times \frac{\left(\frac{S}{2}\right)^{\frac{v}{2}}}{\Gamma\left(\frac{v}{2}\right)(\sigma^2)^{1+\frac{v}{2}}} \exp\left(-\frac{S}{2\sigma^2}\right).
\end{aligned}$$

This has no known form and it is not easy to simulate values of it, so numerical methods are required to explore it. One way to simulate values of this distribution is by means of the Gibbs sampler method, which consists of alternatingly generating samples of the full conditional distributions of each variable (or block of variables) given the rest of the parameters (Casella and George 1992).

The full conditional posteriors to implement the Gibbs sampler are obtained in the lines below.

The conditional posterior distribution of $\boldsymbol{\beta}_0$ is given by

$$
\begin{aligned}
f(\boldsymbol{\beta}_0|-) &\propto L(\mu,\boldsymbol{\beta}_0^{\mathrm{T}},\sigma^2;\boldsymbol{y})f\left(\boldsymbol{\beta}_0|\sigma_\beta^2\right)\\
&\propto \exp\left[-\frac{1}{2\sigma^2}\|\boldsymbol{y}-1_n\mu-\boldsymbol{X}_1\boldsymbol{\beta}_0\|^2-\frac{1}{2\sigma_\beta^2}\boldsymbol{\beta}_0^{\mathrm{T}}\boldsymbol{\beta}_0\right]\\
&\propto \exp\left\{-\frac{1}{2}\left[\boldsymbol{\beta}_0^{\mathrm{T}}\left(\sigma_\beta^{-2}\boldsymbol{I}_n+\sigma^{-2}\boldsymbol{X}_1^{\mathrm{T}}\boldsymbol{X}_1\right)\boldsymbol{\beta}_0-2\sigma^{-2}(\boldsymbol{y}-1_n\mu)^{\mathrm{T}}\boldsymbol{X}_1\boldsymbol{\beta}_0\right]\right\}\\
&\propto \exp\left\{-\frac{1}{2}\left[\left(\boldsymbol{\beta}_0-\widetilde{\boldsymbol{\beta}}_0\right)^{\mathrm{T}}\widetilde{\boldsymbol{\Sigma}}_0^{-1}\left(\boldsymbol{\beta}_0-\widetilde{\boldsymbol{\beta}}_0\right)\right]\right\}
\end{aligned}
$$

$\widetilde{\boldsymbol{\Sigma}}_0=\left(\sigma_\beta^{-2}\boldsymbol{I}_p+\sigma^{-2}\boldsymbol{X}_1^{\mathrm{T}}\boldsymbol{X}_1\right)^{-1}$ and $\widetilde{\boldsymbol{\beta}}_0=\sigma^{-2}\widetilde{\boldsymbol{\Sigma}}_0\boldsymbol{X}_1^{\mathrm{T}}(\boldsymbol{y}-\mathbf{1}_n\mu)$. That is, $\boldsymbol{\beta}_0\mid-\sim N_p\left(\widetilde{\boldsymbol{\beta}}_0,\widetilde{\boldsymbol{\Sigma}}_0\right)$. Similarly, the conditional distribution of μ is $\mu\mid-\sim N(\widetilde{\mu},\widetilde{\sigma}_0^2)$, where $\widetilde{\sigma}_0^2=\frac{\sigma^2}{n}$ and $\widetilde{\mu}=\frac{1}{n}\mathbf{1}_n^{\mathrm{T}}(\boldsymbol{y}-\boldsymbol{X}_1\boldsymbol{\beta}_0)$.

The conditional distribution of σ^2 is

$$
\begin{aligned}
f(\sigma^2|-) &\propto L(\mu,\boldsymbol{\beta}_0^{\mathrm{T}},\sigma^2;\boldsymbol{y})f(\sigma^2)\\
&\propto \frac{1}{(\sigma^2)^{\frac{n}{2}}}\exp\left[-\frac{1}{2\sigma^2}\|\boldsymbol{y}-1_n\mu-\boldsymbol{X}_1\boldsymbol{\beta}_0\|^2\right]\frac{\left(\frac{S}{2}\right)^{\frac{v}{2}}}{\Gamma\left(\frac{v}{2}\right)(\sigma^2)^{1+\frac{v}{2}}}\exp\left(-\frac{S}{2\sigma^2}\right)\\
&\propto \frac{\left(\frac{\widetilde{S}}{2}\right)^{\frac{\widetilde{v}}{2}}}{(\sigma^2)^{1+\frac{\widetilde{v}}{2}}}\exp\left(-\frac{\widetilde{S}}{2\sigma^2}\right),
\end{aligned}
$$

where $\widetilde{v}=v+n$ and $\widetilde{S}=S+\|\boldsymbol{y}-\mathbf{1}_n\mu-\boldsymbol{X}_1\boldsymbol{\beta}_0\|^2$. So $\sigma^2\mid-\sim\chi^{-2}_{\widetilde{v},\widetilde{S}}$, where $\chi^{-2}_{v,s}$ denotes a scaled inverse Chi-squared distribution with parameters v and S. Similarly, $\sigma_\beta^2\mid-\sim\chi^{-2}_{\widetilde{v}_\beta,\widetilde{S}_\beta}$, where $\widetilde{v}_\beta=v_\beta+p$ and $\widetilde{S}_\beta=S_\beta+\boldsymbol{\beta}_0^{\mathrm{T}}\boldsymbol{\beta}_0$.

In summary, for the Ridge regression model, a Gibbs sampler consists of the following steps:

1. Choose initial values for μ, $\boldsymbol{\beta}_0$, and σ^2.
2. Simulate a value of the full conditional distribution of σ_β^2:

$$\sigma_\beta^2\mid\mu,\boldsymbol{\beta}_0,\sigma^2\sim\chi^{-2}_{\widetilde{v},\widetilde{S}_\beta},$$

where $\chi^{-2}_{\widetilde{v},\widetilde{S}_\beta}$ denotes a scaled inverse Chi-square distribution with shape parameter $\widetilde{v}_\beta=v_\beta+p$ and scale parameter $\widetilde{S}_\beta=S_\beta+\boldsymbol{\beta}_0^{\mathrm{T}}\boldsymbol{\beta}_0$.

3. Simulate the full conditional posterior distribution of β_0:

$$\boldsymbol{\beta}_0 \mid \mu, \sigma_\beta^2, \sigma^2 \sim N_p\left(\widetilde{\boldsymbol{\beta}}_0, \widetilde{\boldsymbol{\Sigma}}_0\right),$$

   where $\widetilde{\boldsymbol{\Sigma}}_0 = \left(\sigma_\beta^{-2}\boldsymbol{I}_p + \sigma^{-2}\boldsymbol{X}_1^{\mathrm{T}}\boldsymbol{X}_1\right)^{-1}$ and $\widetilde{\boldsymbol{\beta}}_0 = \sigma^{-2}\widetilde{\boldsymbol{\Sigma}}_0\boldsymbol{X}_1^{\mathrm{T}}(\boldsymbol{y} - \mathbf{1}_n\mu)$
4. Simulate the full conditional distribution of μ:

$$\mu \mid \boldsymbol{\beta}_0, \sigma_\beta^2, \sigma^2 \sim N\left(\widetilde{\mu}, \widetilde{\sigma}_\mu^2\right),$$

   where $\widetilde{\sigma}_\mu^2 = \frac{\sigma^2}{n}$ and $\widetilde{\mu} = \mathbf{1}_n^{\mathrm{T}}(\boldsymbol{y} - \boldsymbol{X}_1\boldsymbol{\beta}_0)$.
5. Simulate the full conditional distribution of σ^2:

$$\sigma^2 \mid \mu, \boldsymbol{\beta}_0, \sigma_\beta^2 \sim \chi^{-2}_{\widetilde{v},\widetilde{S}},$$

   where $\widetilde{v} = v + n$ and $\widetilde{S} = S + \|\boldsymbol{y} - \mathbf{1}_n\mu - \boldsymbol{X}_1\boldsymbol{\beta}_0\|^2$.
6. Repeat steps 2–5 depending on how many values of the parameter vector $(\boldsymbol{\beta}^{\mathrm{T}}, \sigma_\beta^2, \sigma^2)$ you wish to simulate. Usually a large number of iterations are needed and an early part of them are discarded, to finally average the rest of each parameter to obtain estimates of them.

The Gibbs sampler described above can be implemented easily with the BGLR R package: if the hyperparameters S-v and S_β-v_β are not specified, by default the BGLR function assigns $v = v_\beta = 5$, and to S and S_β assigns values such that the mode of the priors of σ^2 and σ_β^2 (inverse scaled Chi-square) matches a certain proportion of the total variance ($1 - R^2$ and R^2): $S = \text{Var}(Y) \times (1 - R^2) \times (v + 2)$ and $S_\beta = \text{Var}(Y) \times R^2 \times (v_\beta + 2)$ (see Appendix 2 for more details). Explicitly, in BGLR this model can be implemented by running the following R code:

```
ETA = list( list( model = 'BRR', X = X₁, df0 = vβ, S0 = Sβ, R2 = 1-R²) )
A = BGLR(y=y, ETA = ETA, nIter = 1e4, burnIn = 1e3, S0 = S, df0 = v, R2 = R²)
```

where nIter = 1e4 and burnIn = 1e3 are the desired number of iterations and the number of them to be discarded when computing the estimates of the parameters. Remember that when the hyperparameter values are not given, they are set up in the default values, as previously described.

A sub-model of the BRR that does not induce shrinkage of the beta coefficients is obtained by assuming that $(\beta_1, \ldots, \beta_p)^{\mathrm{T}} \mid \sigma_\beta^2 \sim N_p\left(\mathbf{0}, \boldsymbol{I}_p\sigma_\beta^2\right)$, ignoring the prior distribution of σ_β^2 and setting this at a very high value (10^{10}). Note that this model is very similar to the Bayesian model obtained by adopting the prior (6.2), under which the beta coefficients are estimated solely with the information contained in the likelihood function (Pérez and de los Campos 2014). This prior model can also be implemented in the BGLR package and is called **FIXED**. Certainly, the Gibbs sampler steps for its implementation are the same as the steps described before for the BRR, except that step 2 is removed (no simulations are obtained from σ_β^2) and σ_β^{-2} is set equal to zero in the full conditional of $\boldsymbol{\beta}_0$ (step 3).

6.3 Bayesian GBLUP Genomic Model

In genomic-enabled prediction, the number of markers used to predict the performance of a trait of interest is often very large compared to the number of individuals phenotyped in the sample ($p \gg n$); for this reason, some computational difficulties may arise when exploring the posterior distribution of the beta coefficients. When the main objective is to use this model for predictive purposes, a solution consists of reducing the dimension of the problem by directly simulating values of $\boldsymbol{g} = \boldsymbol{X}_1\boldsymbol{\beta}_0$ (breeding values or genomic effects, Lehermeier et al. 2013) instead of only from $\boldsymbol{\beta}_0$. To do this, first note that because $\boldsymbol{\beta}_0 \mid \sigma_\beta^2 \sim N_p\left(\mathbf{0}, \boldsymbol{I}_p\sigma_\beta^2\right)$, to induce a prior for $\boldsymbol{g}$, this is defined as $\boldsymbol{g} = \boldsymbol{X}_1\boldsymbol{\beta}_0 \mid \sigma_\beta^2 \sim N_n\left(\mathbf{0}, \sigma_\beta^2\boldsymbol{X}_1\boldsymbol{X}_1^{\mathrm{T}}\right) = N_n\left(\mathbf{0}, \sigma_g^2\boldsymbol{G}\right)$, where $\sigma_g^2 = p\sigma_\beta^2$ and $\boldsymbol{G} = \frac{1}{p}\boldsymbol{X}_1\boldsymbol{X}_1^{\mathrm{T}}$, which is known as the genomic relationship matrix (VanRaden 2007). Then, under this parameterization ($\boldsymbol{g} = \boldsymbol{X}_1\boldsymbol{\beta}_0$ and $\sigma_g^2 = p\sigma_\beta^2$), the model specified in (6.3), in matrix notation takes the following form:

$$\boldsymbol{Y} = 1_n\mu + \boldsymbol{g} + \epsilon \tag{6.4}$$

with a flat prior to mean parameter (μ), $\sigma^2 \sim \chi_{v,S}^{-2}$, and the induced priors: $\boldsymbol{g} = \boldsymbol{X}_1\boldsymbol{\beta}_0 \mid \sigma_g^2 \sim N_n\left(\mathbf{0}, \sigma_g^2\boldsymbol{G}\right)$ and $\sigma_g^2 \sim \chi_{v_g,\ S_g}^{-2}$ ($v_g = v_\beta$, $S_g = pS_\beta$).

Similarly to what was done for model (6.3), the full conditional posterior distribution of $\boldsymbol{g}$ in model (6.4) is given by

$$\begin{aligned}
f(\boldsymbol{g}|-) &\propto L\left(\mu, \boldsymbol{g}, \sigma^2; \boldsymbol{y}\right) f\left(\boldsymbol{g}|\sigma_g^2\right) \\
&\propto \frac{1}{\left(2\pi\sigma^2\right)^{\frac{n}{2}}} \exp\left[-\frac{1}{2\sigma^2}\|\boldsymbol{y} - 1_n\mu - \boldsymbol{g}\|^2\right] \frac{1}{\left(\sigma_g^2\right)^{\frac{n}{2}}} \exp\left(\left[-\frac{1}{2\sigma_g^2}\boldsymbol{g}^{\mathrm{T}}\boldsymbol{G}^{-1}\boldsymbol{g}\right]\right) \\
&\propto \exp\left\{-\frac{1}{2}\left[(\boldsymbol{g} - \widetilde{\boldsymbol{g}})^{\mathrm{T}}\widetilde{\boldsymbol{G}}^{-1}(\boldsymbol{g} - \widetilde{\boldsymbol{g}})\right]\right\},
\end{aligned}$$

where $\widetilde{\boldsymbol{G}} = \left(\sigma_g^{-2}\boldsymbol{G}^{-1} + \sigma^{-2}\boldsymbol{I}_n\right)^{-1}$ and $\widetilde{\boldsymbol{g}} = \sigma^{-2}\widetilde{\boldsymbol{G}}(\boldsymbol{y} - \mathbf{1}_n\mu)$, and from here $\boldsymbol{g} \mid - \sim N_n\left(\widetilde{\boldsymbol{g}}, \widetilde{\boldsymbol{G}}\right)$. Then, the mean/mode of $\boldsymbol{g} \mid -$ is $\widetilde{\boldsymbol{g}} = \sigma^{-2}\widetilde{\boldsymbol{G}}(\boldsymbol{y} - \mathbf{1}_n\mu)$, which is also the best linear unbiased predictor (BLUP) of $\boldsymbol{g}$ under the mixed model equation of Henderson (1975) using the machinery of a classic linear mixed model described in the previous chapter for model (6.4), after recognizing the prior distribution of $\boldsymbol{g}$ as the distribution of the random effects in a model, ignoring the priors specification of the rest of the parameters and assuming that they are known (Henderson 1975). For this reason, model (6.4) is often referred to as GBLUP. If $\boldsymbol{G}$ is replaced by the pedigree matrix $\boldsymbol{A}$, the resulting model is known as PBLUP or ABLUP.

The full conditional posterior of the rest of parameters is similar to the BRR model: $\mu \mid - \sim N(\widetilde{\mu}, \widetilde{\sigma}_0^2)$, where $\widetilde{\sigma}_0^2 = \frac{\sigma^2}{n}$ and $\widetilde{\mu} = \frac{1}{n}\mathbf{1}_n^{\mathrm{T}}(\boldsymbol{y} - \boldsymbol{g})$; $\sigma^2 \mid - \sim \chi^{-2}_{\widetilde{v},\widetilde{S}}$, where $\widetilde{v} = v + n$ and $\widetilde{S} = S + \|\boldsymbol{y} - \mathbf{1}_n\mu - \boldsymbol{g}\|^2$; and $\sigma_g^2 \mid - \sim \chi^{-2}_{\widetilde{v}_g,\widetilde{S}_g}$, where $\widetilde{v}_g = v_g + n$ and $\widetilde{S}_g = S_g + \boldsymbol{g}^{\mathrm{T}}\boldsymbol{G}^{-1}\boldsymbol{g}$.

Note that when $p \gg n$, then the dimension of the parameter space of the posterior of GBLUP model is lower than the BRR.

The GBLUP model (6.4) also can be implemented easily with the BGLR R package, and when the hyperparameters S-v and S_g-v_g are not specified, $v = v_g = 5$ is used by default and the scale parameters are settled similarly as in the BRR.

The BGLR code to fit this model:

```
ETA = list( list( model = 'RHKS', K = G, df0 = vg, S0 = Sg, R2 = 1-R²)) )
A = BGLR(y=y, ETA = ETA, nIter = 1e4, burnIn = 1e3, S0 = S, df0 = v, R2 = R²)
```

The GBLUP can be equivalently expressed and consequently fitted with the BRR model by making the design matrix equal to the lower triangular factor of the Cholesky decomposition of the genomic relationship matrix, i.e., $\boldsymbol{X} = \boldsymbol{L}$, where $\boldsymbol{G} = \boldsymbol{L}\boldsymbol{L}'$. So, with the BGLR package, the BRR implementation of a GBLUP model is.

```
L = t(chol(G))
ETA = list( list( model = 'BRR', X = L, df0 = vβ, S0 = Sβ, R2 = 1-R²) )
A = BGLR(y=y, ETA = ETA, nIter = 1e4, burnIn = 1e3, S0 = S, df0 = v, R2 = R²)
```

When there is more than one repetition of an individual in the data at hand, or a more sophisticated design is used in the data collection, model (6.4) can be specified in a more general way to take into account this structure, as follows:

$$\boldsymbol{Y} = \mathbf{1}_n\mu + \boldsymbol{Z}\boldsymbol{g} + \boldsymbol{\epsilon} \tag{6.5}$$

with $\boldsymbol{Z}$ the incident matrix of the genotypes. This model cannot be fitted directly in the BGLR and some precalculus is needed first to compute the "covariance" matrix of the predictor $\boldsymbol{Zg}$ in model (6.5): $\mathbf{K}_L = \mathrm{Var}(\boldsymbol{Zg}) = \boldsymbol{ZGZ}^{\mathrm{T}}$. The BGLR code for implementing this model is the following:

```
Z = model.matrix(~0+GID,data=dat_F,xlev = list(GID=unique
(dat_F$GID)))
K_L = Z%*%G%*%t(Z)
ETA = list( list( model = 'RHKS', K = K_L, df0 = vg, S0 = Sg, R2 = 1-R²)) )
A = BGLR(y=y, ETA = ETA, nIter = 1e4, burnIn = 1e3, S0 = S, df0 = v, R2 = R²)
```

where **dat_F** is the data set that contains the necessary phenotypic information (GID: Lines or individuals; y: response variable of the trait of interest).

6.4 Genomic-Enabled Prediction BayesA Model

Another variant to the standard Bayesian model (6.1) is the BayesA model proposed by Meuwissen et al. (2001), which is a slight modification of the BRR model obtained with the same prior distributions, except that now a specific variance $\sigma^2_{\beta_j}$ is assumed for each of the covariate (marker) effects, that is, $\beta_j \mid \sigma^2_{\beta_j} \sim N\left(0, \sigma^2_{\beta j}\right)$, and these variance parameters are supposed to be independent random variables with a scaled inverse Chi-square distribution with parameters v_β and S_β, $\sigma^2_{\beta_j} \sim \chi^{-2}(v_\beta, S_\beta)$. These specific variances for each marker effect provide covariate heterogeneous shrinkage estimation. Furthermore, a gamma distribution is assigned to S_β, $S_\beta \sim G(r, s)$, where $G(r, s)$ denotes a gamma distribution where r and s are the rate and shape parameters, respectively. By providing a different prior variance for each β_j, this model has the potential of inducing covariate-specific shrinkage of estimated effects (Pérez and de los Campos 2013).

Note that choosing $r = r^*/v_\beta$ and taking very large values of v_β, the prior of $\sigma^2_{\beta j}$ collapses to a degenerate distribution at S_β, and the BRR model is obtained, but with a gamma distribution with parameters r^* and s as priors to the common variance of the effects $\sigma^2_\beta = \mathrm{Var}(\beta_j) = S_\beta$, instead of $\chi^{-2}(v_\beta, S_\beta)$. Furthermore, the marginal distribution of each beta coefficient β_j, that is, the unconditional distribution of $\beta_j \mid S_\beta$, is a scaled-t-student distribution (scaled by $\sqrt{S_\beta / v_\beta}$). These distributions, compared to the normal, have heavier tails and put higher mass around 0, which compared to the BRR, induce fewer shrinkage estimates of covariates with sizable effects, and induce strong shrinkage toward zero estimates of covariates with smaller effects, respectively (de los Campos et al. 2013).

A Gibbs sampler implementation for estimating the parameters of this model can be done following steps 1–6 of the BRR model, where the second step is replaced by the next 2.1 and 2.2 steps, the third and the last step are replaced and modified by the next steps 3 and 6:

2.1 Simulate from the full conditional of each $\sigma_{\beta^2_j}$

$$\sigma^2_{\beta_j} \mid \mu, \boldsymbol{\beta}_0, \boldsymbol{\sigma}^2_{-j}, S_\beta, \sigma^2 \sim \chi^{-2}_{\widetilde{v}_j, \widetilde{S}_{\beta j}},$$

where $\widetilde{v}_{\beta_j} = v_\beta + 1$ and scale parameter $\widetilde{S}_{\beta_j} = S_\beta + \beta^2_j$, where $\boldsymbol{\sigma}^2_{-j}$ is the vector $\boldsymbol{\sigma}^2_\beta = \left(\sigma^2_{\beta_1}, \ldots, \sigma^2_{\beta_p}\right)$ but without the jth entry.

2.2 Simulate from the full conditional of S_β

$$f(S_\beta|-) \propto \left[\prod_{j=1}^{p} f\left(\sigma^2_{\beta_j}|S_\beta\right)\right] f(S_\beta)$$
$$\propto \prod_{j=1}^{p}\left[\frac{\left(\frac{S_\beta}{2}\right)^{\frac{v_\beta}{2}}}{\Gamma\left(\frac{v_\beta}{2}\right)\left(\sigma^2_{\beta_j}\right)^{1+\frac{v_\beta}{2}}}\exp\left(-\frac{S_\beta}{2\sigma^2_{\beta_j}}\right)\right] S_\beta^{s-1}\exp(-rS_\beta)$$
$$\propto S_\beta^{s+\frac{pv_\beta}{2}-1}\exp\left[-\left(r+\frac{1}{2}\sum_{j=1}^{p}\frac{1}{\sigma^2_{\beta j}}\right)S_\beta\right]$$

which corresponds to the kernel of the gamma distribution with rate parameter $\widetilde{r} = r + \frac{1}{2}\sum_{j=1}^{p}\frac{1}{\sigma^2_{\beta j}}$ and shape parameter $\widetilde{s} = s + \frac{pv_\beta}{2}$, and so $S_\beta \mid - \sim \text{Gamma}(\widetilde{r}, \widetilde{s})$.

3. Simulate the full conditional posterior distribution of $\boldsymbol{\beta}_0$:

$$\boldsymbol{\beta}_0 \mid \mu, \sigma^2_\beta, \sigma^2 \sim N_p\left(\widetilde{\boldsymbol{\beta}}_0, \widetilde{\boldsymbol{\Sigma}}_0\right),$$

where $\widetilde{\boldsymbol{\Sigma}}_0 = \left(\boldsymbol{D}_p^{-1} + \sigma^{-2}\boldsymbol{X}_1^T\boldsymbol{X}_1\right)^{-1}$ and $\widetilde{\boldsymbol{\beta}}_0 = \sigma^{-2}\widetilde{\boldsymbol{\Sigma}}_0\boldsymbol{X}_1^{\mathrm{T}}(\boldsymbol{y} - \boldsymbol{1}_n\mu)$, $\boldsymbol{D}_p = \text{Diag}\left(\sigma^2_{\beta_1}, \ldots, \sigma^2_{\beta_p}\right)$.

6. Repeat steps 2–5 (given in the BRR method) depending on how many values of the parameter vector $(\boldsymbol{\beta}^{\mathrm{T}}, \sigma^2_\beta, \sigma^2, S_\beta)$ we wish to simulate.

When implementing this model in the BGLR package, by default $v = v_\beta = 5$ are used and $S = \text{Var}(Y) \times (1 - R^2) \times (v + 2)$, which makes the mode of the priors of σ^2 ($\chi^{-2}(v, S)$) match a certain proportion of the total variance ($1 - R^2$). If the hyperparameters of the a priori for S_β, s, and r, are not specified, by default BGLR takes $s = 1.1$ to have an a priori ($S_\beta \sim G(s, r)$) that is relatively non-informative with a coefficient of variation ($1/\sqrt{s}$) of approximately 95%. Then, to the rate parameter it assigns the value $r = (s - 1)/S_\beta$, with $S_\beta = \text{Var}(y) \times R^2 \times (v_\beta + 2)/S_x^2$, where S_x^2 is the sum of the variances of the columns of $\boldsymbol{X}$ and R^2 is the percentage of the total variation that a priori is required to attribute to the covariates in $\boldsymbol{X}$. The BGLR code for implementing this model is

```
ETA = list(list(model = 'BayesA', X=X1, df0 = vβ, rate0 = r, shape0 = s,
R2 = 1-R²))
A   = BGLR(y=y, ETA = ETA, nIter = 1e4, burnIn = 1e3, S0 = S, df0 = v, R2 =
R²)
```

6.5 Genomic-Enabled Prediction BayesB and BayesC Models

Other variants of the model (6.1) are the BayesC and the BayesB models, which in turn can be considered as direct extensions of the BRR and BayesA models, respectively, by adding a parameter π that represents the prior proportion of covariates with nonzero effect (Pérez and de los Campos 2014).

The **BayesC** is the same as the BRR, but instead of assuming a priori that the beta coefficients are independent normal random variables with mean 0 and variance σ_β^2, it assumes that with probability π each β_j comes from a $N\left(0, \sigma_\beta^2\right)$, and with probability $1 - \pi$ comes from a degenerate distribution (DG) at zero, that is, $\beta_1, \ldots, \beta_p \mid \sigma_\beta^2, \pi \overset{iid}{\sim} \pi_p \mathrm{N}\left(0, \sigma_\beta^2\right) + \left(1 - \pi_p\right)\mathrm{DG}(0)$ (mix of a normal distribution with mean 0 and variance σ_β^2, and degenerate distribution at zero). In addition, for the parameter π_p, a beta distribution is assigned as prior, that is, $\pi_p \sim \mathrm{Beta}\ (\pi_{p0}, \phi_0)$, where $\pi_{p0} = E(\pi_p)$ represents the mean and ϕ_0^{-1} is the "dispersion" parameter $\left(\mathrm{var}\left(\pi_p\right) = \frac{\pi_{p0}\left(1-\pi_{p0}\right)}{\phi_0+1}\right)$. If $\phi_0 = 2$ and $\pi_{p0} = 0.5$, the prior for π_p is a uniform distribution in (0,1). For large values of ϕ_0, the distribution for π_p is highly concentrated around π_{p0}, and so the BayesC is reduced to BRR when $\pi_{p0} = 1$ for large values of ϕ_0.

For this model, the full conditional distributions of μ and σ_β^2 are the same as the model described before, that is, $\mu \mid - \sim N\left(\widetilde{\mu}, \widetilde{\sigma}_\mu^2\right)$ and $\sigma^2 \mid - \sim \chi_{\widetilde{v},\widetilde{S}}^{-2}$. However, for the rest of the parameters, this does not have a known form and is not easy to simulate from them. A solution is to introduce a latent variable to represent the prior distribution of each β_j, and compute all the conditional distributions in this augmented scheme, including the distribution corresponding to the latent variable. To do this, note that this prior can be specified by assuming that conditional to a binary latent variable Z_j,

$$\beta_j \mid \sigma_\beta^2, Z_j = z \sim \begin{cases} N\left(0, \sigma_\beta^2\right), \\ \mathrm{DG}(0), \end{cases}$$

where Z_j is a Bernoulli random variable with parameter π_p ($Z_j \sim \mathrm{Ber}(\pi_p)$). With this introduced latent variable, all the full conditionals can be derived, as is described next.

If the current value of z_j is 1, the full conditional posterior of β_j is

$$
\begin{aligned}
f(\beta_j|-) &\propto L(\mu, \boldsymbol{\beta}_0, \sigma^2; \boldsymbol{y}) f\left(\beta_j|\sigma_\beta^2, z_j\right) \\
&\propto \exp\left(-\frac{1}{2\sigma^2}\|\boldsymbol{y} - 1_n\mu - \boldsymbol{X}_1\boldsymbol{\beta}_0\|\right) \frac{1}{\sqrt{2\pi\sigma_\beta^2}} \exp\left(-\frac{\beta_j^2}{2\sigma_\beta^2}\right) \\
&\propto \exp\left(-\frac{1}{2\sigma^2}\sum_{i=1}^{n}\left(y_{ij} - x_{ij}\beta_j\right)^2 - \frac{1}{2\sigma_\beta^2}\beta_j^2\right) \\
&\propto \exp\left\{-\frac{1}{2}\left[\left(\sigma_\beta^{-2} + \sigma^{-2}\sum_{i=1}^{n}x_{ij}^2\right)\beta_j^2 - 2\sigma^{-2}\sum_{i=1}^{n}x_{ij}y_{ij}\beta_j + \frac{1}{\sigma^2}\sum_{i=1}^{n}y_{ij}^2\right]\right\} \\
&\propto \exp\left[-\frac{\left(\beta_j - \widetilde{\beta}_j\right)^2}{2\widetilde{\sigma}_j^2}\right],
\end{aligned}
$$

where $y_{ij} = y_i - \sum_{k=1, k \neq j}^{p} x_{ik}\beta_k$, $\widetilde{\sigma}_j^2 = \left(\sigma_\beta^{-2} + \sigma^{-2}\sum_{i=1}^{n}x_{ij}^2\right)^{-1}$, and $\widetilde{\beta}_j = \sigma^{-2}\widetilde{\sigma}_j^2\sum_{i=1}^{n}x_{ij}y_{ij}$. That is, when the current value of z_j is 1, $\beta_j \mid - \sim N\left(\widetilde{\beta}_j, \widetilde{\sigma}_j^2\right)$. However, if $z_j = 0$, the full conditional posterior of β_j is a degenerate random variable at 0, that is, $\beta_j \mid - \sim \text{DG}(0)$.

The full conditional distribution of Z_j is

$$
\begin{aligned}
f(z_j|-) &\propto f\left(\beta_j|\sigma_\beta^2, z_j\right) f(z_j) \\
&\propto f\left(\beta_j|\sigma_\beta^2, z_j\right) \pi_p^{z_j}(1 - \pi_j)^{1-z_j}
\end{aligned}
$$

from which, conditional on the rest of the parameters, Z_j is a Bernoulli random variable with parameter $\widetilde{\pi}_{pj} = \frac{\frac{\pi_p}{\sqrt{2\pi\sigma_\beta^2}}\exp\left(-\frac{\beta_j^2}{2\sigma_\beta^2}\right)}{\frac{\pi_p}{\sqrt{2\pi\sigma_\beta^2}}\exp\left(-\frac{\beta_j^2}{2\sigma_\beta^2}\right) + (1-\pi_p)\delta_0(\beta_j)}$. Note however that, if $\beta_j \neq 0$, $\widetilde{\pi}_{pj} = 1$, and then $Z_j = 1$ with probability 1, when simulating from the full conditional posterior of β_j, we will always obtain values different from zero, and this cyclic behavior will remain permanent. On the other hand, note that if $\beta_j = 0$, $\widetilde{\pi}_{pj} = \frac{\frac{\pi_p}{\sqrt{2\pi\sigma_\beta^2}}}{\frac{\pi_p}{\sqrt{2\pi\sigma_\beta^2}} + (1-\pi_p)}$ is not 0, then the next simulated value of β_j will be different from 0, and in this scenario, in the next steps of the "Gibbs," Z_j will at all times be 1, and so the chain has absorbing states and will not explore the entire sampling space. A solution to this problem consists of trying to simulate from the joint conditional distribution of β_j and Z_j, that is, from $\beta_j, Z_j \mid -$. This full joint conditional distribution can be computed as

$$f(\beta_j, z_j|-) \propto f_*(z_j) f(\beta_j|z_j, -) \propto f_*(z_j) L(\mu, \boldsymbol{\beta}_0, \sigma^2; \mathbf{y}) f\left(\beta_j|\sigma_\beta^2, z_j\right),$$

where $f_*(z_j)$ is the marginal conditional distribution of Z_j conditioned to all parameters except β_j ($Z_j \mid -j \sim f_*(\cdot)$). Specifically, this is given by "integrating" ($f(\beta_j, z_j| -)$ with respect to β_j)

$$\begin{aligned}
f_*(z_j) &\propto \int_{-\infty}^{\infty} f(\beta_j, z_j|-) d\beta_j \propto \int_{-\infty}^{\infty} L(\mu, \boldsymbol{\beta}_0, \sigma^2; \mathbf{y}) f\left(\beta_j|\sigma_\beta^2, z_j\right) f(z_j|\pi_p) d\beta_j \\
&\propto \begin{cases} \int_{-\infty}^{\infty} \exp\left[-\frac{1}{2\sigma^2}\sum_{i=1}^{n}(y_{ij} - x_{ij}\beta_j)^2\right] \frac{1}{\sqrt{2\pi\sigma_\beta^2}} \exp\left(-\frac{\beta_j^2}{2\sigma_\beta^2}\right) \pi_p d\beta_j \\ \int_{-\infty}^{\infty} \exp\left[-\frac{1}{2\sigma^2}\sum_{i=1}^{n}(y_{ij} - x_{ij}\beta_j)^2\right] \delta_0(\beta_j)(1 - \pi_p) d\beta_j \end{cases} \\
&\propto \begin{cases} \pi_p \sqrt{\frac{\widetilde{\sigma}_j^2}{\sigma_\beta^2}} \\ 1 - \pi_p. \end{cases}
\end{aligned}$$

From here, $Z_j \mid -$ is a Bernoulli random distribution with parameter $\widetilde{\pi}_p = \left(\pi_p \sqrt{\frac{\widetilde{\sigma}_j^2}{\sigma_\beta^2}}\right) / \left(\pi_p \sqrt{\frac{\widetilde{\sigma}_j^2}{\sigma_\beta^2}} + 1 - \pi_p\right)$. With this and the full conditional distribution derived above for β_j, an easy way to simulate values from β_j, $Z_j \mid -$ consists of first simulating a z_j value from $Z_j \mid -j \sim \text{Ber}(\widetilde{\pi}_p)$, and then, if $z_j = 1$, simulating a value of β_j from $\beta_j \mid - \sim N\left(\widetilde{\beta}_j, \widetilde{\sigma}_j^2\right)$, otherwise take $\beta_j = 0$.

Now, note that the full conditional distribution of σ_β^2 is

$$\begin{aligned}
f\left(\sigma_\beta^2|-\right) &\propto \left\{\prod_j^p \left[f\left(\beta_j|\sigma_\beta^2, z_j\right)\right]^{z_j}\right\} f\left(\sigma_\beta^2\right) \\
&\propto \frac{1}{\left(\sigma_\beta^2\right)^{p\bar{z}_p}} \exp\left(-\frac{1}{2\sigma_\beta^2}\sum_{j=1}^{p} z_j\beta_j^2\right) \frac{\left(\frac{S_\beta}{2}\right)^{\frac{v_\beta}{2}}}{\Gamma\left(\frac{v_\beta}{2}\right)\left(\sigma_\beta^2\right)^{1+\frac{v_\beta}{2}}} \exp\left(-\frac{S_\beta}{2\sigma_\beta^2}\right) \\
&\propto \frac{1}{\left(\sigma_\beta^2\right)^{p\bar{z}_p}} \exp\left(-\frac{1}{2\sigma_\beta^2}\sum_{j=1}^{p} z_j\beta_j^2\right) \frac{\left(\frac{S_\beta}{2}\right)^{\frac{v_\beta}{2}}}{\Gamma\left(\frac{v_\beta}{2}\right)\left(\sigma_\beta^2\right)^{1+\frac{v_\beta}{2}}} \exp\left(-\frac{S_\beta}{2\sigma_\beta^2}\right) \\
&\propto \frac{1}{\left(\sigma_\beta^2\right)^{1+\frac{\widetilde{v}_\beta}{2}}} \exp\left(-\frac{\widetilde{S}_\beta}{2\sigma_\beta^2}\right),
\end{aligned}$$

where $\bar{z}_p = 1/p\sum_{j=1}^{p} z_j$, $\widetilde{S}_\beta = S_\beta + \sum_{j=1}^{p} z_j\beta_j^2$, and $\widetilde{v}_\beta = v_\beta + p\bar{z}_p$. That is, $\sigma_\beta^2 \mid \mu, \boldsymbol{\beta}_0, \sigma^2, \mathbf{z} \sim \chi^{-2}_{\widetilde{v}, \widetilde{S}_\beta}$. The full conditional distributions of μ and σ^2 are the same as BRR, that is, $\mu \mid - \sim N\left(\widetilde{\mu}, \widetilde{\sigma}_\mu^2\right)$ and $\sigma^2 \mid - \sim \chi^{-2}_{\widetilde{v}, \widetilde{S}}$, with $\widetilde{\sigma}_\mu^2 = \frac{\sigma^2}{n}$, $\widetilde{\mu} = \mathbf{1}_n^{\mathrm{T}}(\boldsymbol{y} - \boldsymbol{X}_1\boldsymbol{\beta}_0)$, $\widetilde{v} = v + n$, and $\widetilde{S} = S + \|\boldsymbol{y} - \mathbf{1}_n\mu - \boldsymbol{X}_1\boldsymbol{\beta}_0\|^2$.

The full conditional distribution of π_p is

$$
\begin{aligned}
f(\pi_p|-) &\propto \left[\prod_{j=1}^{p} f(z_j|\pi_p)\right] f(\pi_p) \\
&\propto \pi_p^{p\bar{z}_p}(1-\pi_p)^{p(1-\bar{z}_p)}\pi_p^{\phi_0\pi_{p0}+1-1}(1-\pi_p)^{\phi_0(1-\pi_{p0})-1} \\
&\propto \pi_p^{\phi_0\pi_{p0}+p\bar{z}_p-1}(1-\pi_p)^{\phi_0(1-\pi_{p0})+p(1-\bar{z}_p)-1}
\end{aligned}
$$

which means that $\pi_p \mid - \sim$ Beta $\left(\widetilde{\pi}_{p0}, \widetilde{\phi}_0\right)$, with $\widetilde{\phi}_0 = \phi_0 + p$ and $\widetilde{\pi}_{p0} = \frac{\phi_0\pi_{p0}+p\bar{z}_p}{\phi_0+p}$.

The **BayesB** model is a variant of BayesA that assumes almost the same prior models to the parameters, except that instead of assuming independent normal random variables with common mean 0 and common variance σ_β^2 for the beta coefficients, this model adopts a mixture distribution, that is, $\beta_j \mid \sigma_{\beta_j}^2, \pi \overset{iid}{\sim} \pi N\left(0, \sigma_{\beta_j}^2\right) + (1-\pi)\mathrm{DG}(0)$, with $\pi \sim \mathrm{Beta}(\pi_0, p_0)$. This model has the potential to perform variable selection and produce covariate-specific shrinkage estimates (Pérez et al. 2010).

This model can also be considered an extension of the BayesC model with a gamma distribution as prior to the scale parameter of the a priori distribution of the variance of the beta coefficients, that is, $S_\beta \sim G(s, r)$. It is interesting to point out that if $\pi = 1$, this model is reduced to BayesA, which is obtained by taking $\pi_0 = 1$ and letting ϕ_0 go to ∞. Also, this is reduced to the BayesC by setting $s/r = S_\beta^0$ and choosing a very large value for r.

To explore the posterior distribution of this model, the same Gibbs sampler given for BayesC can be used, but adding to the process the full conditional posterior distribution of S_β: $S_\beta \mid - \sim \mathrm{Gamma}(\widetilde{r}, \widetilde{s})$, where $\widetilde{r} = r + \frac{1}{2\sigma_\beta^2}$ and shape parameter $\widetilde{s} = s + \frac{v_\beta}{2}$.

When implementing both these models in the BGLR R package, by default this assigns $\pi_{p0} = 0.5$ and $\phi_0 = 10$, for the hyperparameters of the prior model of π_p, Beta (π_{p0}, ϕ_0), which results in a weakly informative prior. For the remaining hyperparameters of the **BayesC** model, by default BGLR assigns values like those assigned to the BRR model, but with some modifications to consider because a priori now only a proportion π_0 of the covariates (columns of $\boldsymbol{X}$) has nonzero effects:

$$
\begin{aligned}
v &= v_\beta = 5,\\
S &= \mathrm{Var}(Y) \times \left(1 - R^2\right) \times (v + 2),\\
S_\beta &= \mathrm{Var}(y) \times R^2 \times \frac{\left(v_\beta + 2\right)}{S_x^2 \pi_0}.
\end{aligned}
$$

While for the remaining hyperparameters of **BayesB**, by default BGLR also assigns values similar to **BayesA**: $v = v_\beta = 5$, $S = \mathrm{Var}(Y) \times (1 - R^2) \times (v + 2)$, $r = (s - 1)/S_\beta$, with $S_\beta = \mathrm{Var}(y) \times R^2 \times \frac{(v_\beta+2)}{S_x^2\pi_0}$, where S_x^2 is the sum of the variances of the columns of $\boldsymbol{X}$.

The BGLR codes to implement these models are, respectively:

```
ETA = list( list( model = 'BayesC', X=X1 ) )
A   = BGLR(y=y, ETA = ETA, nIter = 1e4, burnIn = 1e3, df0 = v, S0 = S, probIn
= πp0, counts = ϕ0, R2 = R²)
```

and

```
ETA = list( list( model = 'BayesB', X=X1 ) )
A    = BGLR(y=y, ETA = ETA, nIter = 1e4, burnIn = 1e3, df0 = v, rate0 = r,
shape0 = s, probIn = πp0, counts = ϕ0, R2 = R²)
```

6.6 Genomic-Enabled Prediction Bayesian Lasso Model

Another variant of the model (6.1) is the Bayesian Lasso linear regression model (**BL**). This model assumes independent Laplace or double-exponential distributions with location and scale parameters 0 and $\frac{\sqrt{\sigma^2}}{\lambda}$, respectively, for the beta coefficients, that is, $\beta_1, \ldots, \beta_p \mid \sigma^2, \lambda \overset{iid}{\sim} L\left(0, \frac{\sqrt{\sigma^2}}{\lambda}\right)$. Furthermore, the priors for parameters μ and σ^2 are the same as in the models described before, while for λ^2, a gamma distribution with parameters s_λ and r_λ is often adopted.

Because compared to the normal distribution, the Laplace distribution has fatter tails and puts higher density around 0, this prior induces stronger shrinkage estimates for covariates with relatively small effects and reduced shrinkage estimates for covariates with larger effects (Pérez et al. 2010).

A more convenient specification of the prior for the beta coefficients in this model is obtained with the representation proposed by Park and Casella (2008), which is a continuous scale mixture of a normal distribution: $\beta_j \mid \tau_j \sim N(0, \tau_j\sigma^2)$ and $\tau_j \sim \mathrm{Exp}(2/\lambda^2)$, $j = 1, \ldots, p$, where $\mathrm{Exp}(\theta)$ denotes an exponential distribution with scale parameter θ. So, unlike the prior used by the BRR model, this prior distribution also puts a higher mass at zero and has heavier tails, which induce stronger shrinkage estimates for covariates with relatively small effect and less shrinkage estimates for markers with sizable effect (Pérez et al. 2010).

Note that the prior distribution for the beta coefficients and the prior variance of this distribution in BayesB and BayesC can be equivalently expressed as a mixture of a scaled inverse Chi-squared distribution with parameters v_β and S_β, and a degenerate distribution at zero, that is, $\beta_j \sim N\left(0, \sigma_\beta^2\right)$ and $\sigma_\beta^2 \sim \pi_p \chi^{-2}(v_\beta, S_\beta) + (1 - \pi_p)\mathrm{DG}(0)$. So, based on this result and the connections between the models described before, the main difference between all these models is the manner in which the prior variance of the predictor variable is modelled.

Example 1 To illustrate how to use the models described before, here we consider the prediction of grain yield (tons/ha) based on marker information. The data set used consists of 30 lines in four environments with one and two repetitions and the genotyped information contains 500 markers for each line. The numbers of lines with one (two) repetition are 6 (24), 2 (28), 0 (30), and 3 (27) in Environments 1, 2, 3, and 4, respectively, resulting in 229 observations. The performance prediction of all these models was evaluated with 10 random partitions in a cross-validation strategy, where 80% of the complete data set was used to fit the model and the rest to evaluate the model in terms of the mean squared error of prediction (MSE).

The results for all models (shown in Table 6.1) were obtained by iterating 10,000 times the corresponding Gibbs sampler and discarding the first 1000 of them, using the default hyperparameter values implemented in BGLR. This indicates that the behavior of all the models is similar, except the BayesC, where the MSE is slightly greater than the rest.

The R code to obtain the results in Table 6.1 is given in Appendix 3.

What happens when using other hyperparameter values? Although the ones used here (proposed by Pérez et al. 2010) did not always produce the best prediction performance (Lehermeier et al. 2013) and there are other ways to propose the hyperparameter values in these models (Habier et al. 2010, 2011), it is important to point out that the values used by default in BGLR work reasonably well and that it

Table 6.1 Mean squared error (MSE) of prediction across 10 random partitions, with 80% for training and the rest for testing, in five Bayesian linear models

PT	BRR	GBLUP	BayesA	BayesB	BayesC
1	0.5395	0.5391	0.5399	0.5415	0.5470
2	0.3351	0.3345	0.3345	0.3411	0.3511
3	0.4044	0.4065	0.4056	0.4050	0.4106
4	0.3540	0.3583	0.3571	0.3561	0.3559
5	0.3564	0.3555	0.3556	0.3549	0.3604
6	0.7781	0.7772	0.7702	0.7777	0.7776
7	0.7430	0.7357	0.7380	0.7384	0.7431
8	0.3662	0.3717	0.3695	0.3669	0.3661
9	0.3065	0.3026	0.3056	0.3021	0.3030
10	0.5842	0.5822	0.5846	0.5860	0.6079
Mean (SD)	0.4767 (0.1742)	0.4763 (0.1724)	0.476 (0.1716)	0.4769 (0.1734)	0.4822 (0.1744)

is not easy to find other combinations that work better in all applications, and when you want to use other combinations of hyperparameters you need to be very careful because you can dramatically affect the predictive performance of the model that uses the default hyperparameters.

Indeed, by means of simulated and experimental data, Lehermeier et al. (2013) observed a strong influence on the predictive performance of the hyperparameters given to the prior distributions in BayesA, BayesB, and the Bayes Lasso with fixed λ. Specifically, in the first two models, they observed that the scale parameter S_β of the prior distribution of variance of β_j had a strong effect on the predictive ability because overfitting in the data occurred when a too large value of this value was chosen, whereas underfitting was observed when too small values of this parameter were used. Note that this is expected approximately by seeing that in both models (BayesA and BayesB), $\mathrm{Var}(\beta_j) = E\left(\sigma^2_{\beta_j}\right) = S_\beta / (v_\beta - 1)$, which is almost the inverse of the regularization parameter in any type of Ridge regression model.

6.7 Extended Predictor in Bayesian Genomic Regression Models

All the Bayesian formulations of the model (6.1) described before can be extended, in terms of the predictor, to easily take into account the effects of other factors. For example, effects of environments and environment–marker interaction can be added as

$$\boldsymbol{y} = \mathbf{1}_n\mu + \boldsymbol{X}_E\boldsymbol{\beta}_E + \boldsymbol{X}\boldsymbol{\beta} + \boldsymbol{X}_{EM}\boldsymbol{\beta}_{EM} + \boldsymbol{\epsilon}, \tag{6.6}$$

where $\boldsymbol{X}_E$ and $\boldsymbol{X}_{EM}$ are the design matrices of the environments and environment–marker interactions, respectively, while $\boldsymbol{\beta}_E$ and $\boldsymbol{\beta}_{EM}$ are the vectors of the environment effects and the interaction effects, respectively, with a prior distribution that can be specified as was done for $\boldsymbol{\beta}$. Indeed, with the BGLR function all these things are possible, and all the options described before can also be adopted for the rest of effects added in the model: FIXED, BRR, BayesA, BayesB, BayesC, and BL.

Under the RKHS model with genotypic and environment–genotypic interaction effects, in the predictor, the modified model (6.6) is expressed as

$$\boldsymbol{Y} = \mathbf{1}_n\mu + \boldsymbol{X}_E\boldsymbol{\beta}_E + \boldsymbol{Z}_L\boldsymbol{g} + \boldsymbol{Z}_{EL}\boldsymbol{gE} + \boldsymbol{\epsilon}, \tag{6.7}$$

where $\boldsymbol{Z}_L$ and $\boldsymbol{Z}_{LE}$ are the incident matrices of the genomic and environment–genotypic interaction effects, respectively. Similarly to model (6.5), this model cannot be fitted directly in the BGLR and some precalculations are needed first to compute the "covariance" matrix of the predictors $\boldsymbol{Z}_L\boldsymbol{g}$ and $\boldsymbol{Z}_{EL}\boldsymbol{gE}$, which are $\boldsymbol{K}_L = \sigma_g^{-2}\,\mathrm{Var}(\boldsymbol{Z}_L\boldsymbol{g}) = \boldsymbol{Z}_L\boldsymbol{G}\boldsymbol{Z}_L^{\mathrm{T}}$ and $\boldsymbol{K}_{LE} = \sigma_{gE}^{-2}\mathrm{Var}(\boldsymbol{Z}_{LE}\boldsymbol{gE}) = \boldsymbol{Z}_{LE}(\boldsymbol{I}_I \otimes \boldsymbol{G})\boldsymbol{Z}_{LE}^{\mathrm{T}}$, respectively, where I is the number of environments. The BGLR code for implementing this model is the following:

```
 I = length(unique(dat_F$Env))
 XE = model.matrix(~0+Env,data=dat_F)[,-1]
 Z_L = model.matrix(~0+GID,data=dat_F,xlev = list(GID=unique
(dat_F$GID)))
 K_L = Z_L%*%G%*%t(Z_L)
 Z_LE = model.matrix(~0+GID:Env,data=dat_F,
        xlev = list(GID=unique(dat_F$GID),Env = unique(dat_F$Env)))
 K_LE = Z_LE%*%kronecker(diag(I),G)%*%t(Z_LE)
    ETA = list( list(model='FIXED',X=XE),
          list( model = 'RHKS', K = K_L, df0 = v_g, S0 = S_g, R2 = 1-R^2)),
            list(model='RKHS',K=K_LE)
A   = BGLR(y,ETA = ETA, nIter = 1e4, burnIn = 1e3, S0 = S, df0 = v, R2 = R^2)
```

where **dat_F** is the data set that contains the necessary phenotypic information (GID: Lines or individuals; Env: Environment; y: response variable of trait under study).

Example 2 (Includes Models with Only Env Effects and Models with Env and LinexEnv Effects) To illustrate how to fit the extended genomic regression models described before, here we consider the prediction of grain yield (tons/ha) based on marker information and the genomic relationship derived from it. The data set used consists of 30 lines in four environments, and the genotyped information of 500 markers for each line. The performance prediction of 18 models was evaluated with a five-fold cross-validation, where 80% of the complete data set was used to fit the model and the rest to evaluate the model in terms of the mean squared error of prediction (MSE). These models were obtained by considering different predictors (marker, environment, or/and environment–marker interaction) and different prior models to the parameters of each predictor included.

The model M1 only considered in the predictor the marker effects, from which six sub-models were obtained by adopting one of the six options (BRR, RKHS, BayesA, BayesB, BayesC, or BL) to the prior model of $\boldsymbol{\beta}$ (or $\boldsymbol{g}$). Model M2 is model M1 plus the environment effects with a FIXED prior model, for all prior sub-models in the marker predictor. Model M3 is model M2 plus the environment–marker interaction, with a prior model of the same family as those chosen for the marker predictor (see in Table 6.2 all the models we compared).

The performance prediction of the models presented in Table 6.2 is shown in Table 6.3. The first column represents the kind of prior model used in both marker effects and env:marker interaction terms, when the latter is included in the model. In each of the first five prior models, model M2 resulted in better MSE performance, while when the BL prior model was used, model M3, the model with the interaction term, was better. The greater difference is between M1 and M2, where the average MSE across all priors of the first model is approximately 21% greater than the corresponding average of the M2 model. Similar behavior was observed with Pearson's correlation, with the average of this criterion across all priors about 32%

Table 6.2 Fitted models: M_{md}, $m = 1, 2, 3, d = 1, \ldots, 6$

Model	Sub-model	Predictor Marker or G: $\boldsymbol{X\beta}$	Environment: $\boldsymbol{X_E\beta_E}$	Interaction Env-Markers: $\boldsymbol{X_{EM}\beta_{EM}}$
M1	M11	BRR		
	M12	RKHS		
	M12	BayesA		
	M14	BayesB		
	M15	BayesC		
	M16	BL		
M2	M21	BRR	FIXED	
	M22	RKHS	FIXED	
	M23	BayesA	FIXED	
	M24	BayesB	FIXED	
	M25	BayesC	FIXED	
	M26	BL	FIXED	
M3	M31	BRR	FIXED	BRR
	M32	RKHS	FIXED	RKHS
	M33	BayesA	FIXED	BayesA
	M34	BayesB	FIXED	BayesB
	M35	BayesC	FIXED	BayesC
	M36	BL	FIXED	BL

The third to fifth columns are the considered predictors corresponding to marker (genetic effect), environment, and environment–marker (genetic effect) effects, respectively. The cells in row 2 and column 3 indicate the prior distribution model specified for the beta parameters in each predictor when that parameter is present. M1 $: \boldsymbol{Y} = \boldsymbol{1}\mu + \boldsymbol{X\beta} + \boldsymbol{\epsilon}$, M2 $: \boldsymbol{Y} = \boldsymbol{1}\mu + \boldsymbol{X_E\beta_E} + \boldsymbol{X\beta} + \boldsymbol{\epsilon}$, and M3 $: \boldsymbol{Y} = \boldsymbol{1}\mu + \boldsymbol{X_E\beta_E} + \boldsymbol{X\beta} + \boldsymbol{X_{EM}\beta_{EM}} + \boldsymbol{\epsilon}$. For the RKHS prior model, predictors $\boldsymbol{X\beta}$ and $\boldsymbol{X_{EM}\beta_{EM}}$ are replaced by $\boldsymbol{g}$ and $\boldsymbol{Eg}$, respectively

greater in model M2 than in M1. So the inclusion of the environment effect was important, but not the environment:marker interaction.

The best prediction performance in terms of MSE was obtained with sub-model M25 (M2 with a BayesC prior) followed by M21 (M2 with a BRR prior). However, the difference between those and sub-models M22, M23, and M24, also derived from M2, is only slight and a little more than with M26, which as commented before, among the models that assume a BL prior, showed a worse performance than M36 (M3 plus a BL prior for marker effects and environment–marker interaction).

6.8 Bayesian Genomic Multi-trait Linear Regression Model

The univariate models described for continuous outcomes do not exploit the possible correlation between traits, when the selection of better individuals is based on several traits and these univariate models are fitted separately to each trait. Relative

Table 6.3 Performance prediction of models in Table 6.2: Mean squared error of prediction (MSE) and average Pearson's correlation (PC), each with its standard deviation across the five partitions

	Markers		Env + Markers		Env + Markers + Env:Markers	
	M1		M2		M3	
Prior	MSE (SD)	PC (SD)	MSE (SD)	PC (SD)	MSE (SD)	PC (SD)
BRR	0.5384 (0.08)	0.4678 (0.11)	0.4446 (0.11)	0.6112 (0.1)	0.4782 (0.12)	0.579 (0.11)
RKHS	0.5509 (0.09)	0.4505 (0.13)	0.4542 (0.12)	0.6012 (0.1)	0.51 (0.13)	0.5271 (0.12)
BayesA	0.5441 (0.08)	0.4622 (0.12)	0.4452 (0.11)	0.616 (0.1)	0.4806 (0.15)	0.5964 (0.11)
BayesB	0.5462 (0.08)	0.4595 (0.12)	0.4455 (0.11)	0.6141 (0.1)	0.4669 (0.13)	0.5981 (0.11)
BayesC	0.5449 (0.08)	0.4615 (0.12)	0.4416 (0.11)	0.614 (0.1)	0.474 (0.12)	0.5864 (0.11)
BL	0.5621 (0.1)	0.4529 (0.12)	0.4719 (0.14)	0.5871 (0.09)	0.4537 (0.12)	0.606 (0.1)

The first column represents the prior used in the model in both marker main effects and env:marker interaction term (when included in the model)

to this, some advantages of jointly modelling the multi-traits is that in this way the correlation among the traits is appropriately accounted for and can help to improve statistical power and the precision of parameter estimation, which are very important in genomic selection, because they can help to improve prediction accuracy and reduce trait selection bias (Schaeffer 1984; Pollak et al. 1984; Montesinos-López et al. 2018).

An example of this is when crop breeders collect phenotypic data for multiple traits such as grain yield and its components (grain type, grain weight, biomass, etc.), tolerance to biotic and abiotic stresses, and grain quality (taste, shape, color, nutrient, and/or content) (Montesinos-López et al. 2016). In this and many other cases, sometimes the interest is to predict traits that are difficult or expensive to measure with those that are easy to measure or the aim can be to improve all these correlated traits simultaneously (Henderson and Quaas 1976; Calus and Veerkamp 2011; Jiang et al. 2015). In these lines, there is evidence of the usefulness of multi-trait modelling. For example, Jia and Jannink (2012) showed that, compared to single-trait modelling, the prediction accuracy of low-heritability traits could be increased by using a multi-trait model when the degree of correlation between traits is at least moderate. They also found that multi-trait models had better prediction accuracy when phenotypes were not available on all individuals and traits. Joint modelling also has been found useful for increasing prediction accuracy when the traits of interest are not measured in the individuals of the testing set, but this and other traits were observed in individuals in the training set (Pszczola et al. 2013).

6.8.1 *Genomic Multi-trait Linear Model*

The genomic multi-trait linear model adopts a univariate genomic linear model structure for each trait but with correlated residuals and genotypic effects for traits in the same individual. Assuming that for individual j, n_T traits (Y_{jt}, $t = 1, \ldots, n_T$) are measured, this model assumes that

$$\begin{bmatrix} Y_{j1} \\ Y_{j2} \\ \vdots \\ Y_{jn_T} \end{bmatrix} = \begin{bmatrix} \mu_1 \\ \mu_2 \\ \vdots \\ \mu_{n_T} \end{bmatrix} + \begin{bmatrix} \boldsymbol{x}_j^{\mathrm{T}}\boldsymbol{\beta}_1 \\ \boldsymbol{x}_j^{\mathrm{T}}\boldsymbol{\beta}_2 \\ \vdots \\ \boldsymbol{x}_j^{\mathrm{T}}\boldsymbol{\beta}_{n_T} \end{bmatrix} + \begin{bmatrix} g_{j1} \\ g_{j2} \\ \vdots \\ g_{jn_T} \end{bmatrix} + \begin{bmatrix} \epsilon_{j1} \\ \epsilon_{j2} \\ \vdots \\ \epsilon_{jn_T} \end{bmatrix},$$

where μ_t, $t = 1, \ldots, n_T$, are the specific trait intercepts, $\boldsymbol{x}_j$ is a vector of covariates equal for all traits, g_{jt}, $t = 1, \ldots, n_T$, are the specific trait genotype effects, and ϵ_{jt}, $t = 1, \ldots, n_T$ are the random error terms corresponding to each trait. In matrix notation, it can be expressed as

$$\boldsymbol{Y}_j = \boldsymbol{\mu} + \boldsymbol{B}^{\mathrm{T}} \boldsymbol{x}_j + \boldsymbol{g}_j + \boldsymbol{\epsilon}_j, \tag{6.8}$$

where $\boldsymbol{Y}_j = \left[Y_{j1}, \ldots, Y_{jn_T}\right]^{\mathrm{T}}$, $\boldsymbol{\mu} = \left[\mu_1, \ldots, \mu_{n_T}\right]^{\mathrm{T}}$, $\boldsymbol{B} = \left[\boldsymbol{\beta}_1, \quad \ldots, \quad \boldsymbol{\beta}_{n_T}\right]$, $\boldsymbol{g}_j = \left[g_{j1}, \ldots, g_{jn_T}\right]^{\mathrm{T}}$, and $\boldsymbol{\epsilon}_j = [\epsilon, \ldots, \epsilon_{n_T}]^{\mathrm{T}}$. The residual vectors are assumed to be independent with multivariate normal distribution, that is $\boldsymbol{\epsilon}_j \sim N_{n_T}(\boldsymbol{0}, \boldsymbol{R})$, and all the random genotype effects are assumed to be $\boldsymbol{g} = \left[\boldsymbol{g}_1^{\mathrm{T}}, \ldots, \boldsymbol{g}_J^{\mathrm{T}}\right]^{\mathrm{T}} \sim N(\boldsymbol{0}, \boldsymbol{G} \otimes \boldsymbol{\Sigma}_{\mathrm{T}})$, $\otimes$ being the Kronecker product. For a full Bayesian specification of this model, we suppose that $\boldsymbol{\beta} = \mathrm{vec}(\boldsymbol{B}) \sim N(\boldsymbol{\beta}_0, \boldsymbol{\Sigma}_{\beta})$, that is, marginally, for the fixed effect of each trait, a prior multivariate normal distribution is adopted, $\boldsymbol{\beta}_t \sim N_p\left(\boldsymbol{\beta}_{t0}, \boldsymbol{\Sigma}_{\boldsymbol{\beta}_t}\right), t = 1, \ldots, n_T$; a flat prior for the intercepts, $f(\boldsymbol{\mu}) \propto 1$; and independent inverse Wishart distributions for the covariance matrix of residuals $\boldsymbol{R}$ and for $\boldsymbol{\Sigma}_T$, that is, $\boldsymbol{\Sigma}_T \sim \mathrm{IW}(v_t, S_t)$ and $\boldsymbol{R} \sim \mathrm{IW}(v_R, S_R)$.

Putting all the information together where the measured traits of each individual ($\boldsymbol{Y}_j$) are accommodated in the rows of a matrix response ($\boldsymbol{Y}$), model (6.8) can be expressed as

$$\boldsymbol{Y} = \boldsymbol{1}_J \boldsymbol{\mu}^{\mathrm{T}} + \boldsymbol{X}\boldsymbol{B} + \boldsymbol{Z}_1 \boldsymbol{b}_1 + \boldsymbol{E}, \tag{6.9}$$

where $=[\boldsymbol{Y}_1, \ldots, \boldsymbol{Y}_J]^{\mathrm{T}}$, $\boldsymbol{X} = [\boldsymbol{x}_1, \ldots, \boldsymbol{x}_J]^{\mathrm{T}}$, $\boldsymbol{b}_1 = [\boldsymbol{g}_1, \ldots, \boldsymbol{g}_J]^{\mathrm{T}}$, and $\boldsymbol{E} = [\boldsymbol{\epsilon}_1, \ldots, \boldsymbol{\epsilon}_J]^{\mathrm{T}}$. Note that under this notation, $\boldsymbol{E}^{\mathrm{T}} \sim MN_{n_T \times J}(\boldsymbol{0}, \boldsymbol{R}, \boldsymbol{I}_J)$ or equivalently $\boldsymbol{E} \sim MN_{J \times n_T}(\boldsymbol{0}, \boldsymbol{I}_J, \boldsymbol{R})$, and $\boldsymbol{b}_1^{\mathrm{T}} \sim MN_{n_T \times J}(\boldsymbol{0}, \boldsymbol{\Sigma}_T, \boldsymbol{G})$ or $\boldsymbol{b}_1 \sim MN_{J \times n_T}(\boldsymbol{0}, \boldsymbol{G}, \boldsymbol{\Sigma}_T)$. Here $\boldsymbol{Z} \sim MN_{J \times n_T}(\boldsymbol{M}, \boldsymbol{U}, \boldsymbol{V})$ means that the random matrix $\boldsymbol{Z}$ follows the matrix variate normal distribution with parameters $\boldsymbol{M}$, $\boldsymbol{U}$, and $\boldsymbol{V}$, or equivalently, that the Jn_T random vector vec($\boldsymbol{Z}$) is distributed as $N_{Jn_T}(\mathrm{vec}(\boldsymbol{M}), \boldsymbol{V} \otimes \boldsymbol{U})$, with vec($\cdot$) denoting the vectorization of a matrix that stacks the columns of this in a single column. Note that when $\boldsymbol{\Sigma}_T$ and $\boldsymbol{R}$ are diagonal matrices, model (6.9) is equivalent to separately fitting a univariate GBLUP model to each trait.

The conditional distribution of all traits is given by

$$\begin{aligned} f(\boldsymbol{Y}|\boldsymbol{\mu},\boldsymbol{\beta},\boldsymbol{b}_1,\boldsymbol{\Sigma}_T,\mathbf{R}) &= \frac{|\boldsymbol{R}|^{-\frac{J}{2}}}{(2\pi)^{Jn_T}} \exp\left\{-\frac{1}{2}\mathrm{tr}\left[\boldsymbol{R}^{-1}\left(\boldsymbol{Y} - 1_J\boldsymbol{\mu}^{\mathrm{T}} - \boldsymbol{X}\boldsymbol{B} - \boldsymbol{Z}_1\boldsymbol{b}_1\right)^{\mathrm{T}}\boldsymbol{I}_J\left(\boldsymbol{Y} - 1_J\boldsymbol{\mu}^{\mathrm{T}} - \boldsymbol{X}\boldsymbol{\beta} - \boldsymbol{Z}_1\boldsymbol{b}_1\right)\right]\right\} \\ &= \frac{|\boldsymbol{R}|^{-\frac{J}{2}}}{(2\pi)^{Jn_T}} \exp\left\{-\frac{1}{2}\left[\mathrm{vec}\left(\boldsymbol{Y} - 1_J\boldsymbol{\mu}^{\mathrm{T}} - \boldsymbol{X}\boldsymbol{B} - \boldsymbol{Z}_1\boldsymbol{b}_1\right)\right]^{\mathrm{T}}\left(\boldsymbol{R}^{-1} \otimes \boldsymbol{I}_J\right)\left[\mathrm{vec}\left(\boldsymbol{Y} - 1_J\boldsymbol{\mu}^{\mathrm{T}} - \boldsymbol{X}\boldsymbol{B} - \boldsymbol{Z}_1\boldsymbol{b}_1\right)\right]\right\} \end{aligned}$$

and the joint posterior of parameters $\boldsymbol{\mu}$, $\boldsymbol{B}$, $\boldsymbol{b}_1$, $\boldsymbol{\Sigma}_T$, and $\boldsymbol{R}$ is given by

$$f(\boldsymbol{\mu}, \boldsymbol{B}, \boldsymbol{b}_1, \boldsymbol{\Sigma}_T, \boldsymbol{R}|\boldsymbol{Y}) \propto f(\boldsymbol{Y}|\boldsymbol{\mu}, \boldsymbol{B}, \boldsymbol{b}_1, \boldsymbol{\Sigma}_T, \mathbf{R}) f(\boldsymbol{b}_1|\boldsymbol{\Sigma}_T) f(\boldsymbol{\Sigma}_T) f(\boldsymbol{\beta}) f(\boldsymbol{R}),$$

where $f(\boldsymbol{b}_1 | \boldsymbol{\Sigma}_T)$ denotes the conditional distribution of the genotype effects, and $f(\boldsymbol{\Sigma}_T)$, $f(\boldsymbol{\beta})$, and $f(\boldsymbol{R})$ denote the prior density distribution of $\boldsymbol{\Sigma}_T$, $\boldsymbol{B}$, and $\boldsymbol{R}$, respectively. This joint posterior distribution of the parameters doesn't have closed form; for this reason, next are derived the full conditional distributions for Gibbs sampler implementation.

Let $\boldsymbol{\beta}_0$ and $\boldsymbol{\Sigma}_\beta$ be the prior mean and variance of $\boldsymbol{\beta} = \text{vec}(\boldsymbol{B})$. Because tr $(\boldsymbol{AB}) = \text{vec}(\boldsymbol{A}^{\mathrm{T}})^{\mathrm{T}}\text{vec}(\boldsymbol{B}) = \text{vec}(\boldsymbol{B})^{\mathrm{T}}\text{vec}(\boldsymbol{A}^{\mathrm{T}})$ and vec $(\boldsymbol{AXB}) = (\boldsymbol{B}^{\mathrm{T}} \otimes \boldsymbol{A})\text{vec}(\boldsymbol{X})$, we have that

$$
\begin{aligned}
P(\boldsymbol{\beta}|-) &\propto f(\boldsymbol{Y}|\boldsymbol{\mu},\boldsymbol{B},\boldsymbol{b}_1,\boldsymbol{\Sigma}_T,\mathbf{R})f(\boldsymbol{\beta}) \\
&\propto \exp\left\{\begin{array}{l} -\frac{1}{2}\left[\text{vec}\left(\boldsymbol{Y}-1_J\boldsymbol{\mu}^{\mathrm{T}}-\boldsymbol{Z}_1\boldsymbol{b}_1\right)-(\boldsymbol{I}_{n_T}\otimes\boldsymbol{X})\text{vec}(\boldsymbol{B})\right]^{\mathrm{T}}(\boldsymbol{R}^{-1}\otimes\boldsymbol{I}_J) \\ \times\left[\text{vec}(\boldsymbol{Y}-1_J\boldsymbol{\mu}^{\mathrm{T}}-\boldsymbol{Z}_1\boldsymbol{b}_1)-(\boldsymbol{I}_{n_T}\otimes\boldsymbol{X})\text{vec}(\boldsymbol{B})\right]-\frac{1}{2}(\boldsymbol{\beta}-\boldsymbol{\beta}_0)^{\mathrm{T}}\boldsymbol{\Sigma}_\beta^{-1}(\boldsymbol{\beta}-\boldsymbol{\beta}_0) \end{array}\right\} \\
&\propto \exp\left\{\begin{array}{l} -\frac{1}{2}\left[\text{vec}\left(\boldsymbol{Y}-1_J\boldsymbol{\mu}^{\mathrm{T}}-\boldsymbol{Z}_1\boldsymbol{b}_1\right)-(\boldsymbol{I}_{n_T}\otimes\boldsymbol{X})\boldsymbol{\beta}\right]^{\mathrm{T}}(\boldsymbol{R}^{-1}\otimes\boldsymbol{I}_J) \\ \times\left[\text{vec}(\boldsymbol{Y}-1_J\boldsymbol{\mu}^{\mathrm{T}}-\boldsymbol{Z}_1\boldsymbol{b}_1)-(\boldsymbol{I}_{n_T}\otimes\boldsymbol{X})\boldsymbol{\beta}\right]-\frac{1}{2}(\boldsymbol{\beta}-\boldsymbol{\beta}_0)^{\mathrm{T}}\boldsymbol{\Sigma}_\beta^{-1}(\boldsymbol{\beta}-\boldsymbol{\beta}_0) \end{array}\right\} \\
&\propto \exp\left\{-\frac{1}{2}\left[\boldsymbol{\beta}-\widetilde{\boldsymbol{\beta}}_0\right]^{\mathrm{T}}\widetilde{\boldsymbol{\Sigma}}_\beta^{-1}\left[\boldsymbol{\beta}-\widetilde{\boldsymbol{\beta}}_0\right]\right\},
\end{aligned}
$$

where $\widetilde{\boldsymbol{\Sigma}}_\beta = \left[\boldsymbol{\Sigma}_\beta^{-1} + \left(\boldsymbol{R}^{-1}\otimes\boldsymbol{X}^{\mathrm{T}}\boldsymbol{X}\right)\right]^{-1}$ and $\widetilde{\boldsymbol{\beta}}_0 = \widetilde{\boldsymbol{\Sigma}}_\beta\left[\boldsymbol{\Sigma}_\beta^{-1}\boldsymbol{\beta}_0 + \left(\boldsymbol{R}^{-1}\otimes\boldsymbol{X}^{\mathrm{T}}\right)\text{vec}\right.$ $\left.(\boldsymbol{Y}-\mathbf{1}_J\boldsymbol{\mu}^{\mathrm{T}}-\boldsymbol{Z}_1\boldsymbol{b}_1)\right]$. So, the full conditional distribution of $\boldsymbol{\beta}$ is $N_p\left(\widetilde{\boldsymbol{\beta}}_0, \widetilde{\boldsymbol{\Sigma}}_\beta\right)$. Similarly, the full conditional distribution of $\boldsymbol{g} = \text{vec}(\boldsymbol{b}_1)$ is $N_J\left(\widetilde{\boldsymbol{g}}, \widetilde{\boldsymbol{G}}\right)$, with $\widetilde{\boldsymbol{G}} = \left[\left(\boldsymbol{\Sigma}_T^{-1}\otimes\boldsymbol{G}^{-1}\right)+\left(\boldsymbol{R}^{-1}\otimes\boldsymbol{Z}_1^{\mathrm{T}}\boldsymbol{Z}_1\right)\right]^{-1}$ and $\widetilde{\boldsymbol{g}} = \widetilde{\boldsymbol{G}}\left(\boldsymbol{R}^{-1}\otimes\boldsymbol{Z}_1^{\mathrm{T}}\right)\text{vec}(\boldsymbol{Y}-\mathbf{1}_J\boldsymbol{\mu}^{\mathrm{T}}-\boldsymbol{XB})$. Now, because $\text{vec}(\mathbf{1}_J\boldsymbol{\mu}^{\mathrm{T}}) = (\boldsymbol{I}_{n_T}\otimes\mathbf{1}_J)\boldsymbol{\mu}$, similarly as before, the full conditional of $\boldsymbol{\mu}$ is $N_{n_T}\left(\widetilde{\boldsymbol{\mu}}, \widetilde{\boldsymbol{\Sigma}}_\mu\right)$, where $\widetilde{\boldsymbol{\Sigma}}_\mu = J^{-1}\boldsymbol{R}$ and $\widetilde{\boldsymbol{\mu}} = \widetilde{\boldsymbol{\Sigma}}_\mu\left(\boldsymbol{R}^{-1}\otimes\mathbf{1}_J\right)\text{vec}(\boldsymbol{Y}-\boldsymbol{XB}-\boldsymbol{Z}_1\boldsymbol{B})$.

The full conditional distribution of $\boldsymbol{\Sigma}_T$

$$
\begin{aligned}
P(\boldsymbol{\Sigma}_T|-) &\propto P(\boldsymbol{b}_1|\boldsymbol{\Sigma}_T)P(\boldsymbol{\Sigma}_T) \\
&\propto |\boldsymbol{\Sigma}_T|^{-\frac{J}{2}}|\boldsymbol{G}|^{-\frac{n_T}{2}}\exp\left\{-\frac{1}{2}\text{tr}\left[\boldsymbol{b}_1^{\mathrm{T}}\boldsymbol{G}^{-1}\boldsymbol{b}_1\boldsymbol{\Sigma}_T^{-1}\right]\right\}P(\boldsymbol{\Sigma}_T) \\
&\propto \boldsymbol{\Sigma}_T^{-\frac{\nu_T+J+n_T+1}{2}}\exp\left\{-\frac{1}{2}\text{tr}\left(\boldsymbol{b}_1^{\mathrm{T}}\boldsymbol{G}^{-1}\boldsymbol{b}_1+\boldsymbol{S}_T\right)\boldsymbol{\Sigma}_T^{-1}\right\}.
\end{aligned}
$$

From here we have that $\boldsymbol{\Sigma}_T \mid - \sim \text{IW}\left(\nu_T+J, \boldsymbol{b}_1^{\mathrm{T}}\boldsymbol{G}^{-1}\boldsymbol{b}_1+\boldsymbol{S}_T\right)$. Now, because

$$
\begin{aligned}
P(\boldsymbol{R}|-) &\propto f(\boldsymbol{Y}|\boldsymbol{\mu},\boldsymbol{B},\boldsymbol{b}_1,\boldsymbol{\Sigma}_T,\mathbf{R})f(\boldsymbol{R}) \\
&\propto |\boldsymbol{R}|^{-\frac{J}{2}}\exp\left\{-\frac{1}{2}\text{tr}\left[\boldsymbol{R}^{-1}(\boldsymbol{Y}-1_J\boldsymbol{\mu}^{\mathrm{T}}-\boldsymbol{XB}-\boldsymbol{Z}_1\boldsymbol{b}_1)^{\mathrm{T}}\boldsymbol{I}_J(\boldsymbol{Y}-1_J\boldsymbol{\mu}^{\mathrm{T}}-\boldsymbol{X\beta}-\boldsymbol{Z}_1\boldsymbol{b}_1)\right]\right\} \\
&\quad |\boldsymbol{R}|^{-\frac{\nu_R+n_T+1}{2}}\exp\left\{-\frac{1}{2}\text{tr}\left(\boldsymbol{S}_T\boldsymbol{R}^{-1}\right)\right\} \\
&\propto |\boldsymbol{R}|^{-\frac{\nu_R+J+n_T+1}{2}}\exp\left\{-\frac{1}{2}\text{tr}\left[\boldsymbol{S}_T+(\boldsymbol{Y}-1_J\boldsymbol{\mu}^{\mathrm{T}}-\boldsymbol{XB}-\boldsymbol{Z}_1\boldsymbol{b}_1)^{\mathrm{T}}(\boldsymbol{Y}-1_J\boldsymbol{\mu}^{\mathrm{T}}-\boldsymbol{X\beta}-\boldsymbol{Z}_1\boldsymbol{b}_1)\right]\boldsymbol{R}^{-1}\right\}
\end{aligned}
$$

the full conditional distribution of $\boldsymbol{R}$ is $\text{IW}\left(\widetilde{\nu}_R, \widetilde{\boldsymbol{S}}_R\right)$, where $\widetilde{\nu}_R = \nu_R + J$ and $\widetilde{\boldsymbol{S}}_R = \boldsymbol{S}_T + (\boldsymbol{Y}-\mathbf{1}_J\boldsymbol{\mu}^{\mathrm{T}}-\boldsymbol{XB}-\boldsymbol{Z}_1\boldsymbol{b}_1)^{\mathrm{T}}(\boldsymbol{Y}-\mathbf{1}_J\boldsymbol{\mu}^{\mathrm{T}}-\boldsymbol{X\beta}-\boldsymbol{Z}_1\boldsymbol{b}_1)$.

In summary, a Gibbs sampler exploration of the joint posterior distribution of $\boldsymbol{\mu}$, $\boldsymbol{\beta}$, $\boldsymbol{g}$, $\boldsymbol{\Sigma}_T$, and $\boldsymbol{R}$ can be done with the following steps:

1. Simulate $\boldsymbol{\beta}$ from a multivariate normal distribution $N_p\left(\widetilde{\boldsymbol{\beta}}_0, \widetilde{\boldsymbol{\Sigma}}_\beta\right)$, where $\widetilde{\boldsymbol{\Sigma}}_\beta = \left[\boldsymbol{\Sigma}_\beta^{-1} + \left(\boldsymbol{R}^{-1} \otimes \boldsymbol{X}^{\mathrm{T}}\boldsymbol{X}\right)\right]^{-1}$ and $\widetilde{\boldsymbol{\beta}}_0 = \widetilde{\boldsymbol{\Sigma}}_\beta\left[\boldsymbol{\Sigma}_\beta^{-1}\boldsymbol{\beta}_0 + \left(\boldsymbol{R}^{-1} \otimes \boldsymbol{X}^{\mathrm{T}}\right)\mathrm{vec}\left(\boldsymbol{Y} - \boldsymbol{1}_J\boldsymbol{\mu}^{\mathrm{T}} - \boldsymbol{Z}_1\boldsymbol{b}_1\right)\right]$.
2. Simulate $\boldsymbol{\mu}$ from $N_{n_T}\left(\widetilde{\boldsymbol{\mu}}, \widetilde{\boldsymbol{\Sigma}}_\mu\right)$, where $\widetilde{\boldsymbol{\Sigma}}_\mu = J^{-1}\boldsymbol{R}$ and $\widetilde{\boldsymbol{\mu}} = \widetilde{\boldsymbol{\Sigma}}_\mu\left(\boldsymbol{R}^{-1} \otimes \boldsymbol{1}_J\right)$ $\mathrm{vec}(\boldsymbol{Y} - \boldsymbol{XB} - \boldsymbol{Z}_1\boldsymbol{B})$.
3. Simulate $\boldsymbol{g} = \mathrm{vec}(\boldsymbol{b}_1)$ from $N_J\left(\widetilde{\boldsymbol{g}}, \widetilde{\boldsymbol{G}}\right)$, where $\widetilde{\boldsymbol{G}} = \left[\left(\boldsymbol{\Sigma}_T^{-1} \otimes \boldsymbol{G}^{-1}\right) + \left(\boldsymbol{R}^{-1} \otimes \boldsymbol{Z}_1^{\mathrm{T}}\boldsymbol{Z}_1\right)\right]^{-1}$ and $\widetilde{\boldsymbol{g}} = \widetilde{\boldsymbol{G}}\left(\boldsymbol{R}^{-1} \otimes \boldsymbol{Z}_1^{\mathrm{T}}\right)\mathrm{vec}\left(\boldsymbol{Y} - \boldsymbol{1}_J\boldsymbol{\mu}^{\mathrm{T}} - \boldsymbol{XB}\right)$.
4. Simulate $\boldsymbol{\Sigma}_T$ from $\mathrm{IW}\left(v_T + J, \boldsymbol{b}_1^{\mathrm{T}}\boldsymbol{G}^{-1}\boldsymbol{b}_1 + \boldsymbol{S}_T\right)$.
5. Simulate $\boldsymbol{R}$ from $\mathrm{IW}\left(\widetilde{v}_R, \widetilde{\boldsymbol{S}}_R\right)$, where $\widetilde{v}_R = v_R + J$ and $\widetilde{\boldsymbol{S}}_R = \boldsymbol{S}_T + \left(\boldsymbol{Y} - \boldsymbol{1}_J\boldsymbol{\mu}^{\mathrm{T}} - \boldsymbol{XB} - \boldsymbol{Z}_1\boldsymbol{b}_1\right)^{\mathrm{T}}\left(\boldsymbol{Y} - \boldsymbol{1}_J\boldsymbol{\mu}^{\mathrm{T}} - \boldsymbol{X\beta} - \boldsymbol{Z}_1\boldsymbol{b}_1\right)$.
6. Return to step 1 or terminate when chain length is adequate to meet convergence diagnostics and the required sample size is reached.

An implementation of this model can be done using the github version of the BGLR R library, which can be accessed from https://github.com/gdlc/BGLR-R and can be installed directly in the R console by running the following commands: *install.packages('devtools'); library(devtools); install_git('https://github.com/gdlc/BGLR-R')*. This implementation also uses a flat prior for the fixed effect regression coefficients $\boldsymbol{\beta}$, and in such a case, the corresponding full conditional of this parameter is the same as step 1 of the Gibbs sampler given before, but removing $\boldsymbol{\Sigma}_\beta^{-1}$ and $\boldsymbol{\Sigma}_\beta^{-1}\boldsymbol{\beta}_0$ from $\widetilde{\boldsymbol{\Sigma}}_\beta$ and $\widetilde{\boldsymbol{\beta}}_0$, respectively. Specifically, model (6.8) can be implemented with this version of the BGLR package as follows:

```
ETA = list( list( X=X, model='FIXED' ), list( K=Z₁GZ₁ᵀ, model=' RKHS' ) )
A   = Multitrait(y = Y, ETA=ETA, resCov = list( type = 'UN', S0 = S_R, df0 =
v_R ), nIter = nI, burnIn = nb)
```

The first argument in the Multitrait function is the response variable which many times is a phenotype matrix where each row corresponds to the measurement of n_T traits in each individual. The second argument is a list predictor in which the first sub-list specifies the design matrix and prior model to the fixed effects part of the predictor in model (6.9), and in the second sub-list are specified the parameters of the distribution of random genetic effects of $\boldsymbol{b}_1$, where it is specified the $K = \boldsymbol{G}$ genomic relationship matrix, that accounts for the similarity between individuals based on marker information, df0 $= v_T$ and $S0 = \boldsymbol{S}_T$ are the degrees of freedom parameter (v_T) and the scale matrix parameter ($\boldsymbol{S}_T$) of the inverse Wishart prior distribution for $\boldsymbol{\Sigma}_T$, respectively. In the third argument (resCOV), $S0$ and df0 are the scale matrix parameter ($\boldsymbol{S}_R$) and the degree of freedom parameter (v_R) of the inverse Wishart prior distribution for $\boldsymbol{R}$. The last two arguments are the required number of iterations (**nI**) and the burn-in period (**nb**) for running the Gibbs sampler.

Similarly to the univariate case, model (6.9) can be equivalently described and implemented as a multivariate Ridge regression model, as follows:

$$\boldsymbol{Y} = \boldsymbol{1}_J\boldsymbol{\mu}^{\mathrm{T}} + \boldsymbol{XB} + \boldsymbol{X}_1\boldsymbol{B}_1 + \boldsymbol{E}, \tag{6.10}$$

where $\boldsymbol{X}_1 = \boldsymbol{Z}_1\boldsymbol{L}_G$, $\boldsymbol{G} = \boldsymbol{L}_G\boldsymbol{L}_G^{\mathrm{T}}$ is the Cholesky factorization of $\boldsymbol{G}$, $\boldsymbol{B}_1 = \boldsymbol{L}_G^{-1}\boldsymbol{b}_1 \sim MN_{J\times n_T}(\boldsymbol{0}, \boldsymbol{I}_J, \boldsymbol{\Sigma}_T)$, and the specifications for the rest of parameters and prior distribution are the same as given in model (6.8). A Gibbs sampler implementation of this model is very similar to the one described before, with little modification. Indeed, a Gibbs implementation with the same multi-trait function is as follows:

```
LG= t(chol(G))
X1= Z1LG
ETA = list( list( X=X, model='FIXED' ), list( X=X1, model='BRR' ) )
A   = Multitrait(y = Y, ETA=ETA, resCov = list( type = 'UN', S0 = SR, df0 =
vR ), nIter = nI, burnIn = nb)
```

with the only change in the second sub-list predictor, where now the design matrix $\boldsymbol{X}_1$ and the Ridge regression model (**BRR**) are specified.

Example 3 To illustrate the performance in terms of the prediction power of these models and how to implement this in R software, we considered a reduced data set that consisted of 50 wheat lines grown in two environments. In each individual, two traits were measured: FLRSDS and MIXTIM. The evaluation was done with a five-fold cross-validation, where lines were evaluated in some environments with all traits but are missing for all traits in other environments. Model (6.9) was fitted and the environment effect was assumed a fixed effect.

The results are shown in Table 6.4, where the average (standard deviation) of two performance criteria is shown for each trait in each environment: average Pearson's correlation (PC) and the mean squared error of prediction (MSE). Table 6.4 shows good correlation performance in both traits and in both environments, and better predictions were obtained in environment 2 with both criteria. The magnitude of the

Table 6.4 Average Pearson's correlation (PC) and mean squared error of prediction (MSE) between predicted and observed values across five random partitions where lines were evaluated in some environments with all traits but are missing for all traits in other environments, SD represent standard deviation across partitions

Env	Trait	PC (SD)	MSE (SD)
1	FLRSDS	0.6252 (0.276)	4.3009 (2.343)
	MIXTIM	0.6709 (0.367)	0.457 (0.343)
2	FLRSDS	0.6895 (0.219)	3.9553 (2.171)
	MIXTIM	0.7523 (0.157)	0.3764 (0.249)

MSE in the first trait is mainly because the measurement scale is greater than in the second trait.

The R codes to reproduce these results (Table 6.4) are shown in Appendix 5.

6.9 Bayesian Genomic Multi-trait and Multi-environment Model (BMTME)

Model (6.9) does not take into account the possible trait–genotype–environment interaction ($T \times G \times E$), when environment information is available. An extension of this model is the one proposed by Montesinos-López et al. (2016), who added this interaction term to vary the specific trait genetic effects ($\boldsymbol{g}_j$) across environments. If the information of n_{T} traits of J lines is collected in I environments, this model is given by

$$\boldsymbol{Y} = \mathbf{1}_{IJ}\boldsymbol{\mu}^{\mathrm{T}} + \boldsymbol{XB} + \boldsymbol{Z}_1\boldsymbol{b}_1 + \boldsymbol{Z}_2\boldsymbol{b}_2 + \boldsymbol{E}, \tag{6.11}$$

where $=[\boldsymbol{Y}_1,\ldots,\boldsymbol{Y}_{IJ}]^{\mathrm{T}}$, $\boldsymbol{X} = [\boldsymbol{x}_1,\ldots,\boldsymbol{x}_{IJ}]^{\mathrm{T}}$, $\boldsymbol{Z}_1$ and $\boldsymbol{Z}_2$ are the incident lines and the incident environment–line interaction matrices, $\boldsymbol{b}_1 = [\boldsymbol{g}_1,\ldots,\boldsymbol{g}_J]^{\mathrm{T}}$, $\boldsymbol{b}_2 = [\boldsymbol{g}_{21},\ldots,\boldsymbol{g}_{2IJ}]^{\mathrm{T}}$, and $\boldsymbol{E} = [\boldsymbol{\epsilon}_1,\ldots,\boldsymbol{\epsilon}_{IJ}]^{\mathrm{T}}$. Here, $\boldsymbol{b}_2 \mid \boldsymbol{\Sigma}_T, \boldsymbol{\Sigma}_E \sim MN_{IJ\times n_T}(\boldsymbol{0}, \boldsymbol{\Sigma}_E \otimes \boldsymbol{G}, \boldsymbol{\Sigma}_T)$, and similar to model (6.2), $\boldsymbol{b}_1 \mid \boldsymbol{\Sigma}_T \sim MN_{J\times n_T}(\boldsymbol{0}, \boldsymbol{G}, \boldsymbol{\Sigma}_T)$ and $\boldsymbol{E} \sim MN_{IJ\times n_T}(\boldsymbol{0}, \boldsymbol{I}_{IJ}, \boldsymbol{R})$. The complete Bayesian specification of this model assumes independent multivariate normal distributions for the columns of $\boldsymbol{B}$, that is, for the fixed effect of each trait a prior multivariate normal distribution is adopted, $\boldsymbol{\beta}_t \sim N_p(\boldsymbol{\beta}_{t0}, \boldsymbol{\Sigma}_{\beta_t}), t = 1,\ldots,n_T$; a flat prior for the intercepts, $f(\boldsymbol{\mu}) \propto 1$; and independent inverse Wishart distributions for the covariance matrices of residuals $\boldsymbol{R}$ and for $\boldsymbol{\Sigma}_T$, $\boldsymbol{\Sigma}_T \sim \mathrm{IW}(v_T, \boldsymbol{S}_T)$ and $\boldsymbol{R} \sim \mathrm{IW}(v_R, \boldsymbol{S}_R)$, and also an inverse Wishart distribution for $\boldsymbol{\Sigma}_E$, $\boldsymbol{\Sigma}_E \sim \mathrm{IW}(v_E, \boldsymbol{S}_E)$.

The full conditional distributions of $\boldsymbol{\mu}$, $\boldsymbol{B}$, $\boldsymbol{b}_1$, $\boldsymbol{b}_2$, and $\boldsymbol{R}$ can be derived as in model (6.9). The full conditional distribution of $\boldsymbol{\Sigma}_T$ is

$$\begin{aligned}
f(\boldsymbol{\Sigma}_T|-) &\propto f(\boldsymbol{b}_1|\boldsymbol{\Sigma}_T)P(\boldsymbol{b}_2|\boldsymbol{\Sigma}_T,\boldsymbol{\Sigma}_E)P(\boldsymbol{\Sigma}_T)\\
&\propto |\boldsymbol{\Sigma}_T|^{-\frac{J}{2}}|\boldsymbol{G}|^{-\frac{L}{2}}\exp\left\{-\frac{1}{2}\mathrm{tr}\left[\boldsymbol{b}_1^{\mathrm{T}}\boldsymbol{G}^{-1}\boldsymbol{b}_1\boldsymbol{\Sigma}_T^{-1}\right]\right\}\\
&\quad\times|\boldsymbol{\Sigma}_T|^{-\frac{IJ}{2}}|\boldsymbol{\Sigma}_E\otimes\boldsymbol{G}|^{-\frac{n_T}{2}}\exp\left\{-\frac{1}{2}\mathrm{tr}\left[\boldsymbol{b}_2^{\mathrm{T}}(\boldsymbol{\Sigma}_E^{-1}\otimes\boldsymbol{G}^{-1})\boldsymbol{b}_2\boldsymbol{\Sigma}_T^{-1}\right]\right\}P(\boldsymbol{\Sigma}_T)\\
&\propto \boldsymbol{\Sigma}_T^{-\frac{v_T+J+IJ+n_T+1}{2}}\exp\left\{-\frac{1}{2}\mathrm{tr}(\boldsymbol{b}_1^{\mathrm{T}}\boldsymbol{G}^{-1}\boldsymbol{b}_1+\boldsymbol{b}_2^{\mathrm{T}}(\boldsymbol{\Sigma}_E^{-1}\otimes\boldsymbol{G}^{-1})\boldsymbol{b}_2+\boldsymbol{S}_t)\boldsymbol{\Sigma}_T^{-1}\right\},
\end{aligned}$$

that is, $\boldsymbol{\Sigma}_T \mid - \sim \mathrm{IW}\left(v_T + J + IJ, \boldsymbol{b}_1^{\mathrm{T}}\boldsymbol{G}^{-1}\boldsymbol{b}_1 + \boldsymbol{b}_2^{\mathrm{T}}(\boldsymbol{\Sigma}_E^{-1}\otimes\boldsymbol{G}^{-1})\boldsymbol{b}_2 + \boldsymbol{S}_T\right)$.

Now, let be $\boldsymbol{b}_2^*$ a $Jn_T \times I$ matrix such that $\text{vec}(\boldsymbol{b}_2^{\text{T}}) = \text{vec}(\boldsymbol{b}_2^*)$. Then because $\boldsymbol{b}_2^{\text{T}} \mid \boldsymbol{\Sigma}_T, \boldsymbol{\Sigma}_E \sim MN_{n_T \times IJ}(\boldsymbol{0}, \boldsymbol{\Sigma}_T, \boldsymbol{\Sigma}_E \otimes \boldsymbol{G})$, $\text{vec}(\boldsymbol{b}_2^{\text{T}}) \mid \boldsymbol{\Sigma}_T, \boldsymbol{\Sigma}_E \sim N(\boldsymbol{0}, \boldsymbol{\Sigma}_E \otimes (\boldsymbol{G} \otimes \boldsymbol{\Sigma}_{\text{T}}))$, and $\boldsymbol{b}_2^* \mid \boldsymbol{\Sigma}_T, \boldsymbol{\Sigma}_E \sim MN_{n_T \times IJ}(\boldsymbol{0}, \boldsymbol{G} \otimes \boldsymbol{\Sigma}_{\text{T}}, \boldsymbol{\Sigma}_E)$, the full conditional posterior distribution of $\boldsymbol{\Sigma}_E$ is

$$\begin{aligned} P(\boldsymbol{\Sigma}_E|\text{ELSE}) &\propto P(\boldsymbol{b}_2|\boldsymbol{\Sigma}_E)\, P(\boldsymbol{\Sigma}_E) \\ &\propto |\boldsymbol{\Sigma}_E|^{-\frac{JL}{2}} |\boldsymbol{G} \otimes \boldsymbol{\Sigma}_T|^{-\frac{I}{2}} \exp\left\{-\frac{1}{2}\text{tr}\left[\boldsymbol{\Sigma}_E^{-1} \boldsymbol{b}_2^{*\text{T}} \left(\boldsymbol{G}^{-1} \otimes \boldsymbol{\Sigma}_T^{-1}\right) \boldsymbol{b}_2^*\right]\right\} \\ &\quad |\boldsymbol{S}_E|^{\frac{\upsilon_E + I - 1}{2}} \times |\boldsymbol{\Sigma}_E|^{-\frac{\upsilon_E}{2}} \exp\left\{-\frac{1}{2}\text{tr}\left(\boldsymbol{S}_E \boldsymbol{\Sigma}_E^{-1}\right)\right\} \\ &\propto |\boldsymbol{\Sigma}_E|^{-\frac{\upsilon_E + JL + I + 1}{2}} \exp\left\{-\frac{1}{2}\text{tr}\left[\left(\boldsymbol{b}_2^{*\text{T}} \left(\boldsymbol{G}^{-1} \otimes \boldsymbol{\Sigma}_T^{-1}\right) \boldsymbol{b}_2^* + \boldsymbol{S}_E\right)\right] \boldsymbol{\Sigma}_E^{-1}\right\} \end{aligned}$$

which means that $\boldsymbol{\Sigma}_E \mid - \sim \text{IW}\left(\upsilon_E + JL, \boldsymbol{b}_2^{*\text{T}} \left(\boldsymbol{G}^{-1} \otimes \boldsymbol{\Sigma}_T^{-1}\right) \boldsymbol{b}_2^* + \boldsymbol{S}_E\right)$.

A Gibbs sampler to explore the joint posterior distribution of parameters of model (6.11), $\boldsymbol{\mu}$, $\boldsymbol{\beta}$, $\boldsymbol{b}_1$,$\boldsymbol{b}_2$, $\boldsymbol{\Sigma}_T$, $\boldsymbol{\Sigma}_E$, and $\boldsymbol{R}$, can be implemented with the following steps:

1. Simulate $\boldsymbol{\beta}$ from a multivariate normal distribution $N_p\left(\widetilde{\boldsymbol{\beta}}_0, \widetilde{\boldsymbol{\Sigma}}_\beta\right)$, where $\widetilde{\boldsymbol{\Sigma}}_\beta = \left[\boldsymbol{\Sigma}_\beta^{-1} + \left(\boldsymbol{R}^{-1} \otimes \boldsymbol{X}^{\text{T}}\boldsymbol{X}\right)\right]^{-1}$ and $\widetilde{\boldsymbol{\beta}}_0 = \widetilde{\boldsymbol{\Sigma}}_\beta\left[\boldsymbol{\Sigma}_\beta^{-1}\boldsymbol{\beta}_0 + \left(\boldsymbol{R}^{-1} \otimes \boldsymbol{X}^{\text{T}}\right)\right.$ $\left.\text{vec}(\boldsymbol{Y} - \boldsymbol{1}_{IJ}\boldsymbol{\mu}^{\text{T}} - \boldsymbol{Z}_1\boldsymbol{b}_1 - \boldsymbol{Z}_2\boldsymbol{b}_2)\right]$.
2. Simulate $\boldsymbol{\mu}$ from $N_{n_T}\left(\widetilde{\boldsymbol{\mu}}, \widetilde{\boldsymbol{\Sigma}}_\mu\right)$, where $\widetilde{\boldsymbol{\Sigma}}_\mu = (IJ)^{-1}\boldsymbol{R}$ and $\widetilde{\boldsymbol{\mu}} = \widetilde{\boldsymbol{\Sigma}}_\mu\left(\boldsymbol{R}^{-1} \otimes \boldsymbol{1}_{IJ}\right)$ $\text{vec}(\boldsymbol{Y} - \boldsymbol{XB} - \boldsymbol{Z}_1\boldsymbol{b}_1 - \boldsymbol{Z}_2\boldsymbol{b}_2)$.
3. Simulate $\boldsymbol{g}_1 = \text{vec}(\boldsymbol{b}_1)$ from $N_J\left(\widetilde{\boldsymbol{g}}_1, \widetilde{\boldsymbol{G}}\right)$, where $\widetilde{\boldsymbol{G}} = \left[\left(\boldsymbol{\Sigma}_T^{-1} \otimes \boldsymbol{G}^{-1}\right) + \left(\boldsymbol{R}^{-1} \otimes \boldsymbol{Z}_1^{\text{T}}\boldsymbol{Z}_1\right)\right]^{-1}$ and $\widetilde{\boldsymbol{g}} = \widetilde{\boldsymbol{G}}\left(\boldsymbol{R}^{-1} \otimes \boldsymbol{Z}_1^{\text{T}}\right)\text{vec}(\boldsymbol{Y} - \boldsymbol{1}_{IJ}\boldsymbol{\mu}^{\text{T}} - \boldsymbol{XB} - \boldsymbol{Z}_2\boldsymbol{b}_2)$.
4. Simulate $\boldsymbol{g}_2 = \text{vec}(\boldsymbol{b}_2)$ from $N_J\left(\widetilde{\boldsymbol{g}}_2, \widetilde{\boldsymbol{G}}_2\right)$, where $\widetilde{\boldsymbol{G}}_2 = \left[\left(\boldsymbol{\Sigma}_T^{-1} \otimes \boldsymbol{\Sigma}_E^{-1} \otimes \boldsymbol{G}^{-1}\right) + \left(\boldsymbol{R}^{-1} \otimes \boldsymbol{Z}_2^{\text{T}}\boldsymbol{Z}_2\right)\right]^{-1}$ and $\widetilde{\boldsymbol{g}}_2 = \widetilde{\boldsymbol{G}}_2\left(\boldsymbol{R}^{-1} \otimes \boldsymbol{Z}_2^{\text{T}}\right)\text{vec}(\boldsymbol{Y} - \boldsymbol{1}_{IJ}\boldsymbol{\mu}^{\text{T}} - \boldsymbol{XB} - \boldsymbol{Z}_1\boldsymbol{b}_1)$.
5. Simulate $\boldsymbol{\Sigma}_T$ from $\text{IW}\left(\nu_T + J + IJ, \boldsymbol{b}_1^{\text{T}}\boldsymbol{G}^{-1}\boldsymbol{b}_1 + \boldsymbol{b}_2^{\text{T}}\left(\boldsymbol{\Sigma}_E^{-1} \otimes \boldsymbol{G}^{-1}\right)\boldsymbol{b}_2 + \boldsymbol{S}_T\right)$.
6. Simulate $\boldsymbol{\Sigma}_E$ from $\text{IW}\left(\upsilon_E + JL, \boldsymbol{b}_2^{*\text{T}}\left(\boldsymbol{G}^{-1} \otimes \boldsymbol{\Sigma}_T^{-1}\right)\boldsymbol{b}_2^* + \boldsymbol{S}_E\right)$.
7. Simulate $\boldsymbol{R}$ from $\text{IW}\left(\widetilde{\nu}_R, \widetilde{\boldsymbol{S}}_R\right)$, where $\widetilde{\nu}_R = \nu_R + IJ$ and $\widetilde{\boldsymbol{S}}_R = \boldsymbol{S}_T + (\boldsymbol{Y} - \boldsymbol{1}_{IJ}\boldsymbol{\mu}^{\text{T}} - \boldsymbol{XB} - \boldsymbol{Z}_1\boldsymbol{b}_1 - \boldsymbol{Z}_2\boldsymbol{b}_2)^{\text{T}}(\boldsymbol{Y} - \boldsymbol{1}_{IJ}\boldsymbol{\mu}^{\text{T}} - \boldsymbol{X\beta} - \boldsymbol{Z}_1\boldsymbol{b}_1 - \boldsymbol{Z}_2\boldsymbol{b}_2)$.
8. Return to step 1 or terminate when chain length is adequate to meet convergence diagnostics and the required sample size is reached.

A similar Gibbs sampler is implemented in the BMTME R package, with the main difference, that this package does not allow specifying a general fixed effect design matrix $\boldsymbol{X}$, only the corresponding to the design matrix for the environment effects, and also the intercept vector $\boldsymbol{\mu}$ is ignored because it is included in the fixed environment effects. Specifically, to fit model (6.11) where the only fixed effects to be taken into account are the environment's effects, the R code to implement this with the BMTME package is as follows:

```
XE    = model.matrix(~Env,data=dat_F)
Z1    = model.matrix(~0+GID,data=dat_F)
Lg    = t(chol(G))
Z1_a = Z1%*%Lg
Z2    = model.matrix(~0+GID:Env,data=dat_F)
G2   = kronecker(diag(dim(XE)[2])),Lg)
Z2 _a = Z2%*%G2
A = BMTME(Y = Y, X = XE, Z1 = Z1_a, Z2 = Z2_a, nIter = nI, burnIn = nb, thin =
2, bs = 50)
```

where **Y** is the matrix of response variables where each row corresponds to the measurement of n_T traits in each individual, **XE** is a design matrix for the environment effects, **Z1** is the incidence matrix of the genetics effects, **Z2** is the design matrix of the genetic–environment interaction effects, **nI** and **nb** are the required number of interactions and the burn-in period, and *bs* is the number of blocks to use internally to sample from vec($\boldsymbol{b}_2$).

Example 4 To illustrate how to implement this model with the BMTME R package, we considered the data in Example 2, but now the explored model includes the trait–genotype–environment interaction.

The average results of the prediction performance in terms of PC and MSE for implementing the same five-fold cross-validation used in Example 3 are shown in Table 6.5. These results show an improvement in terms of prediction performance with this model in all trait environments combinations and in both criteria (PC and MSE) to measure the prediction performance, except in trait MIXTIM and Env 2, where the MSE is slightly greater than the one obtained with model (6.9), which does not take into account the triple interaction (trait–genotype–environment).

The R code used to obtained these results is given in Appendix 5.

Table 6.5 Average Pearson's correlation (PC) and mean squared error of prediction (MSE) between predicted and observed values across five random partitions where lines were evaluated in some environments with all traits but are missing for all traits in other environments

Env	Trait	PC (SD)	MSE (SD)
1	FLRSDS	0.668 (0.243)	3.4483 (1.9)
	MIXTIM	0.6913 (0.35)	0.4255 (0.325)
2	FLRSDS	0.7218 (0.219)	3.2304 (2.034)
	MIXTIM	0.7667 (0.188)	0.3806 (0.287)

Appendix 1

- The probability density function (pdf) of the scaled inverse Chi-square distribution with v degrees of freedom and scale parameter S, $\chi^{-2}(v, S)$, is given by

$$f(\sigma^2; v, S) = \frac{\left(\frac{S}{2}\right)^{\frac{v}{2}}}{\Gamma\left(\frac{v}{2}\right)\left(\sigma^2\right)^{1+\frac{v}{2}}} \exp\left(-\frac{S}{2\sigma^2}\right).$$

and the mean, mode, and variance of this distribution are given by $\frac{S}{v-2}$, $\frac{S}{v+2}$, and $\frac{2S^2}{(v-2)^2(v-4)}$, respectively.

- The pdf of the gamma distribution with shape parameter s and rate parameter r:

$$f_{S_\beta}(x; s, r) = \frac{r^s x^{s-1}}{\Gamma(s)} \exp(-rx).$$

The mean, mode, and variance of this distribution are s/r, $(s - 1)/r$, and s/r^2, respectively.

- The pdf of a beta distribution with mean μ and precision parameter ϕ ("dispersion" parameter ϕ^{-1}) is given by

$$f(x; \mu, \phi) = \frac{1}{B[\mu\phi, (1-\mu)\phi]} x^{\mu\phi-1}(1-x)^{(1-\mu)\phi-1},$$

where the relation with the standard parameterization of this distribution, Beta (α, β), is

$$\mu = \frac{\alpha}{\alpha+\beta}, \phi = \alpha + \beta.$$

- The pdf of a Laplace distribution is

$$f(x; b) = \frac{1}{2b} \exp\left(-\frac{|x|}{b}\right), b > 0, -\infty \le x \le \infty.$$

The mean and variance of this distribution are 0 and $2b^2$.

- A random matrix $\mathbf{\Sigma}$ of dimension $p \times p$ is distributed as inverse Wishart distribution with parameter v and $\mathbf{S}$, $\mathbf{\Sigma} \sim \text{IW}(v, \mathbf{S})$, if it has a density function

$$f(\mathbf{\Sigma}) = \frac{1}{2^{\frac{vp}{2}}\Gamma_p(v/2)} |\mathbf{S}|^{\frac{v}{2}} |\mathbf{\Sigma}|^{-\frac{v+p+1}{2}} \exp\left[-\frac{1}{2}\text{tr}\left(\mathbf{S}\mathbf{\Sigma}^{-1}\right)\right],$$

where $\Gamma_p(v/2)$ is the multivariate gamma function, $v > 0$, and $\mathbf{\Sigma}$ and $\mathbf{S}$ are positive defined matrices. The mean matrix of this distribution is $\frac{\mathbf{S}}{v-p-1}$.

- A $(p \times q)$ random matrix $\boldsymbol{Z}$ follows the matrix normal distribution with matrix parameters $\boldsymbol{M}$ $(p \times q)$, $\boldsymbol{U}$ $(p \times p)$, and $\boldsymbol{V}(q \times q)$, $\boldsymbol{Z} \sim MN_{p,\,q}(\boldsymbol{M}, \boldsymbol{U}, \boldsymbol{V})$, if it has density

$$f(\boldsymbol{Z}|\boldsymbol{M}, \boldsymbol{U}, \boldsymbol{V}) = \frac{\exp\left\{-\frac{1}{2}\mathrm{tr}\left[\boldsymbol{V}^{-1}(\boldsymbol{Z}-\boldsymbol{M})^{\mathrm{T}}\boldsymbol{U}^{-1}(\boldsymbol{Z}-\boldsymbol{M})\right]\right\}}{(2\pi)^{pq/2}|\boldsymbol{V}|^{p/2}|\boldsymbol{U}|^{q/2}}.$$

Appendix 2: Setting Hyperparameters for the Prior Distributions of the BRR Model

The following rules are those used in Pérez and de los Campos (2014), and provide proper but weakly informative prior distributions. In general, this consists of assigning a certain proportion of the total variance of the phenotypes, to the different components of the model.

Specifically, for model (6.3), first the total variance of y is partitioned into two components: (1) the error and (2) the linear predictor:

$$\mathrm{Var}(y_j) = \mathrm{Var}\left(\boldsymbol{x}_j^{\mathrm{T}}\boldsymbol{\beta}_0\right) + \sigma^2$$

Therefore, the average of the variance of the individuals, called total variance, is equal to

$$\frac{1}{n}\sum_{j=1}^{n}\mathrm{Var}(y_j) = \frac{1}{n}\sum_{j=1}^{n}\mathrm{Var}\left(\boldsymbol{x}_j^{\mathrm{T}}\boldsymbol{\beta}_0\right) + \sigma^2 = \frac{1}{n}\mathrm{tr}(\boldsymbol{X}\boldsymbol{X}^{\mathrm{T}})\sigma_\beta^2 + \sigma^2 = V_M + V_\epsilon.$$

Then, by setting R_1^2 as the proportion of the total variance ($\mathbf{V}_y$), that is explained by markers a priori, $V_M = R_1^2\mathbf{V}_y$, and replacing σ_β^2 in V_M by its prior mode, $\frac{S_\beta}{v_\beta+2}$, we have that

$$\frac{1}{n}\mathrm{tr}(\boldsymbol{X}\boldsymbol{X}^{\mathrm{T}})\left(\frac{S_\beta}{v_\beta+2}\right) = R_1^2\mathbf{V}_y.$$

From here, once we have set a value for v_β, the scale parameter is given by

$$S_\beta = \frac{R_1^2\mathbf{V}_y}{\frac{1}{n}\mathrm{tr}(\boldsymbol{X}\boldsymbol{X}^{\mathrm{T}})}(v_\beta + 2).$$

A commonly used value of the shape parameter is $\nu_\beta = 5$ and the value for the proportion of explained variances is $R_1^2 = 0.5$.

Because the model only has two predictors and R_1^2 was set as the proportion of the total variance that is explained by markers a priori, the corresponding proportion that is explained by error a priori is $R_2^2 = 1 - R_1^2$. Then, similar to what was done before, once there is a value for the shape parameter of the prior distribution of σ^2, ν, the value of the scale parameter is given by

$$S = \left(1 - R_1^2\right)\mathbf{V}_y(\nu + 2).$$

By default, $\nu = 5$ is often used.

Appendix 3: R Code Example 1

```
rm(list=ls())
library(BGLR)
load('dat_ls_E1.RData',verbose=TRUE)
#Phenotypic data
dat_F = dat_ls$dat_F
head(dat_F)
#Marker data
dat_M = dat_ls$dat_M
dim(dat_M)

dat_F = transform(dat_F, GID = as.character(GID))
head(dat_F,5)

#Matrix design of markers
Pos = match(dat_F$GID,row.names(dat_M))
XM = dat_M[Pos,]
XM = scale(XM)
dim(XM)

n = dim(dat_F)[1]
y = dat_F$y

#10 random partitions
K = 10
set.seed(1)
PT = replicate(K,sample(n,0.20*n))

#BRR
ETA_BRR = list(list(model='BRR',X=XM))
Tab = data.frame(PT = 1:K,MSEP = NA)
set.seed(1)
for(k in 1:K)
{
```

```
 Pos_tst = PT[,k]
 y_NA = y
 y_NA[Pos_tst] = NA
 A = BGLR(y=y_NA,ETA=ETA_BRR,nIter = 1e4,burnIn = 1e3,verbose = FALSE)
 yp_ts = A$yHat[Pos_tst]
 Tab$MSEP[k] = mean((y[Pos_tst]-yp_ts)^2)
}

#GBLUP
dat_M = scale(dat_M)
G = tcrossprod(XM)/dim(XM)[2]
dim(G)
#Matrix design of GIDs
Z = model.matrix(~0+GID,data=dat_F,xlev = list(GID=unique
(dat_F$GID)))
K_L = Z%*%G%*%t(Z)
ETA_GB = list(list(model='RKHS',K = K_L))
#Tab = data.frame(PT = 1:K,MSEP = NA)
set.seed(1)
for(k in 1:K)
{
 Pos_tst = PT[,k]
 y_NA = y
 y_NA[Pos_tst] = NA
 A = BGLR(y=y_NA,ETA=ETA_GB,nIter = 1e4,burnIn = 1e3,verbose = FALSE)
 yp_ts = A$yHat[Pos_tst]
 Tab$MSEP_GB[k] = mean((y[Pos_tst]-yp_ts)^2)
}

#BA
ETA_BA = list(list(model='BayesA',X=XM))
#Tab = data.frame(PT = 1:K,MSEP = NA)
set.seed(1)
for(k in 1:K)
{
 Pos_tst = PT[,k]
 y_NA = y
 y_NA[Pos_tst] = NA
 A = BGLR(y=y_NA,ETA=ETA_BA,nIter = 1e4,burnIn = 1e3,verbose = FALSE)
 yp_ts = A$yHat[Pos_tst]
 Tab$MSEP_BA[k] = mean((y[Pos_tst]-yp_ts)^2)
}

#BB
ETA_BB = list(list(model='BayesB',X=XM))
#Tab = data.frame(PT = 1:K,MSEP = NA)
set.seed(1)
for(k in 1:K)
{
 Pos_tst = PT[,k]
 y_NA = y
 y_NA[Pos_tst] = NA
 A = BGLR(y=y_NA,ETA=ETA_BB,nIter = 1e4,burnIn = 1e3,verbose = FALSE)
```

```
 yp_ts = A$yHat[Pos_tst]
 Tab$MSEP_BB[k] = mean((y[Pos_tst]-yp_ts)^2)
}

#BC
ETA_BC = list(list(model='BayesC',X=XM))
#Tab = data.frame(PT = 1:K,MSEP = NA)
set.seed(1)
for(k in 1:K)
{
 Pos_tst = PT[,k]
 y_NA = y
 y_NA[Pos_tst] = NA
 A = BGLR(y=y_NA,ETA=ETA_BC,nIter = 1e4,burnIn = 1e3,verbose = FALSE)
 yp_ts = A$yHat[Pos_tst]
 Tab$MSEP_BC[k] = mean((y[Pos_tst]-yp_ts)^2)
}

#BL
ETA_BL = list(list(model='BL',X=XM))
#Tab = data.frame(PT = 1:K,MSEP = NA)
set.seed(1)
for(k in 1:K)
{
 Pos_tst = PT[,k]
 y_NA = y
 y_NA[Pos_tst] = NA
 A = BGLR(y=y_NA,ETA=ETA_BL,nIter = 1e4,burnIn = 1e3,verbose = FALSE)
 yp_ts = A$yHat[Pos_tst]
 Tab$MSEP_BL[k] = mean((y[Pos_tst]-yp_ts)^2)
}
Tab

#Mean and SD across the five partitions
apply(Tab[,-1],2,function(x)c(mean(x),sd(x)))
```

Appendix 4: R Code Example 2

```
rm(list=ls())
library(BGLR)
library(BMTME)
load('dat_ls_E2.RData',verbose=TRUE)
#Phenotypic data
dat_F = dat_ls$dat_F
head(dat_F)
dim(dat_F)
#Marker data
dat_M = dat_ls$dat_M
dim(dat_M)
```

```
dat_F = transform(dat_F, GID = as.character(GID))
head(dat_F,5)

#Matrix design for markers
Pos = match(dat_F$GID,row.names(dat_M))
XM = dat_M[Pos,]
dim(XM)
XM = scale(XM)
#Environment design matrix
XE = model.matrix(~0+Env,data=dat_F)[,-1]
head(XE)
#Environment-marker design matrix
XEM = model.matrix(~0+XM:XE)
#GID design matrix and Environment-GID design matrix
#for RKHS models
Z_L = model.matrix(~0+GID,data=dat_F,xlev = list(GID=unique
(dat_F$GID)))
Z_LE = model.matrix(~0+GID:Env,data=dat_F,
        xlev = list(GID=unique(dat_F$GID),Env = unique(dat_F$Env)))
#Genomic relationship matrix derived from markers
dat_M = scale(dat_M)
G = tcrossprod(dat_M)/dim(dat_M)[2]
dim(G)
#Covariance matrix for Zg
K_L = Z_L%*%G%*%t(Z_L)
#Covariance matrix for random effects ZEg
K_LE = Z_LE%*%kronecker(diag(4),G)%*%t(Z_LE)
n = dim(dat_F)[1]
y = dat_F$y

#Number of random partitions
K = 5
PT = CV.KFold(dat_F,DataSetID = 'GID', K=5, set_seed = 1)
Models = c('BRR','RKHS','BayesA','BayesB','BayesC','BL')
Tab = data.frame()
for(m in 1:6)
{
 ETA1 = list(list(model=Models[m],X=XM))
 ETA2 = list(list(model='FIXED',X=XE),list(model=Models[m],X=XM))
 ETA3 = list(list(model='FIXED',X=XE),list(model=Models[m],X=XM),
       list(model=Models[m],X=XEM))
 if(Models[m]=='RKHS')
 {
  ETA1 = list(list(model='RKHS',K=K_L))
  ETA2 = list(list(model='FIXED',X=XE),list(model='RKHS',K=K_L))
  ETA3 = list(list(model='FIXED',X=XE),list(model='RKHS',K=K_L),
       list(model='RKHS',K=K_LE))
 }
 Tab1_m = data.frame(PT = 1:K,MSEP = NA)
 set.seed(1)
 Tab2_m = Tab1_m
 Tab3_m = Tab2_m
 for(k in 1:K)
```

```
  {
   Pos_tst = PT$CrossValidation_list[[k]]
   y_NA = y
   y_NA[Pos_tst] = NA
   A = BGLR(y=y_NA,ETA=ETA1,nIter = 1e4,burnIn = 1e3,verbose = FALSE)
   yp_ts = A$yHat[Pos_tst]
   Tab1_m$MSEP[k] = mean((y[Pos_tst]-yp_ts)^2)
   Tab1_m$Cor[k]  = cor(y[Pos_tst],yp_ts)

   A = BGLR(y=y_NA,ETA=ETA2,nIter = 1e4,burnIn = 1e3,verbose = FALSE)
   yp_ts = A$yHat[Pos_tst]
   Tab2_m$MSEP[k] = mean((y[Pos_tst]-yp_ts)^2)
   Tab2_m$Cor[k]  = cor(y[Pos_tst],yp_ts)

   A = BGLR(y=y_NA,ETA=ETA3,nIter = 1e4,burnIn = 1e3,verbose = FALSE)
   yp_ts = A$yHat[Pos_tst]
   Tab3_m$MSEP[k] = mean((y[Pos_tst]-yp_ts)^2)
   Tab3_m$Cor[k]  = cor(y[Pos_tst],yp_ts)
  }
 Tab = rbind(Tab,data.frame(Model=Models[m],Tab1_m,Tab2_m,Tab3_m))
}
Tab
```

Appendix 5

R Code Example 3

```
rm(list=ls(all=TRUE))
library(BGLR)
library(BMTME)
library(dplyr)

load('dat_ls.RData',verbose=TRUE)
dat_F = dat_ls$dat_F
head(dat_F)
Y = as.matrix(dat_F[,-(1:2)])
dat_F$Env = as.character(dat_F$DS)
G = dat_ls$G
J = dim(G)[1]
XE = matrix(model.matrix(~0+Env,data=dat_F)[,-1],nc=1)
Z = model.matrix(~0+GID,data=dat_F)
K = Z%*%G%*%t(Z)
#Partitions for a 5-FCV
PT_ls = CV.KFold(dat_FF, DataSetID='GID',K=5,set_seed = 123)
PT_ls = PT_ls$CrossValidation_list
#Predictor BGLR
ETA = list(list(X=XE,model='FIXED'),list(K=K,model='RKHS'))
#Function to summarize the performance prediction: PC_MM_f
source('PC_MM.R')#See below
Tab = data.frame()
```

Appendix 5

```
set.seed(1)
for(p in 1:5)
{
 Y_NA = Y
 Pos_NA = PT_ls[[p]]
 Y_NA[Pos_NA,] = NA
 A = Multitrait(y = Y_NA, ETA=ETA,
          resCov = list( type = 'UN', S0 = diag(2), df0 = 5 ),
          nIter = 5e3, burnIn = 1e3)
 PC = PC_MM_f(Y[Pos_NA,],A$yHat[Pos_NA,],Env=dat_F$Env[Pos_NA])
 Tab = rbind(Tab,data.frame(PT=p,PC))
 cat('PT=',p,'\n')
}
Tab_R = Tab%>%group_by(Env,Trait)%>%select(Cor,MSEP)%>%summarise
(Cor_mean = mean(Cor),
                                       Cor_sd  = sd(Cor),
                                       MSEP_mean = mean(MSEP),
                                       MSEP_sd = sd(MSEP))
Tab_R = as.data.frame(Tab_R)
Tab_R
#Save in the same folder the following in a file with name "PC_MM.R"
#Performance criteria
PC_MM_f<-function(y,yp,Env=NULL)
{
 if(is.null(Env))
 {
  Cor  = diag(cor(as.matrix(y),as.matrix(yp)))
  MSEP = colMeans((y-yp)^2)
  PC = data.frame(Trait = colnames(y),Cor=Cor, MSEP=MSEP)
 }
 else
 {
  PC = data.frame()
  Envs = unique(Env)
  nE = length(Envs)
  for(e in 1:nE)
  {
   y_e = y[Env==Envs[e],]
   yp_e = yp[Env==Envs[e],]
   Cor  = diag(cor(as.matrix(y_e),as.matrix(yp_e)))
   MSEP = colMeans((y_e-yp_e)^2)
   PC = rbind(PC,data.frame(Trait = colnames(y),Env=Envs[e],Cor=Cor,
MSEP=MSEP))
  }
 }
 PC
}
```

R Code for Example 4

```
rm(list=ls(all=TRUE))
library(BMTME)
library(dplyr)
load('dat_ls.RData',verbose=TRUE)
dat_F = dat_ls$dat_F
head(dat_F)
Y = as.matrix(dat_F[,-(1:2)])
dat_F$Env = as.character(dat_F$DS)
G = dat_ls$G
Lg = t(chol(G))
XE = model.matrix(~Env,data=dat_F)
Z1 = model.matrix(~0+GID,data=dat_F)
Z1_a = Z1%*%Lg
Z2 = model.matrix(~0+GID:Env,data=dat_F)
L2 = kronecker(diag(dim(XE)[2]),Lg)
Z2_a = Z2%*%L2
#Partitions for a 5-FCV
PT_ls = CV.KFold(dat_FF, DataSetID='GID',K=5,set_seed = 123)
PT_ls = PT_ls$CrossValidation_list
source('PC_MM.R')#See file R "PC_MM.R" defined in example 6.1
Tab = data.frame()
set.seed(1)
for(p in 1:5)
{
 Y_NA = Y
 Pos_NA = PT_ls[[p]]
 Y_NA[Pos_NA,] = NA
 A = BMTME(Y = Y_NA, X = XE, Z1 = Z1_a, Z2 = Z2_a,
          nIter = 3e3, burnIn = 5e2, thin = 2, bs = 50)
 PC = PC_MM_f(Y[Pos_NA,],A$yHat[Pos_NA,],Env=dat_F$Env[Pos_NA])
 Tab = rbind(Tab,data.frame(PT=p,PC))
 cat('PT=',p,'\n')
}
Tab_R = Tab%>%group_by(Env,Trait)%>%select(Cor,MSEP)%>%summarise
(Cor_mean = mean(Cor),
                              Cor_sd  = sd(Cor),
                              MSEP_mean = mean(MSEP),
                              MSEP_sd = sd(MSEP))
Tab_R = as.data.frame(Tab_R)
Tab_R
```

References

Box GEP, Tiao GC (1992) Bayesian inference in statistical analysis. Wiley, New York

Calus MP, Veerkamp RF (2011) Accuracy of multi-trait genomic selection using different methods. Genet Sel Evol 43(1):26

Casella G, George EI (1992) Explaining the Gibbs sampler. Am Stat 46(3):167–174

Christensen R, Johnson W, Branscum A, Hanson TE (2011) Bayesian ideas and data analysis: an introduction for scientists and statisticians. Chapman & Hall/CRC, Stanford, CA

de los Campos G, Hickey JM, Pong-Wong R, Daetwyler HD, Calus MP (2013) Whole-genome regression and prediction methods applied to plant and animal breeding. Genetics 193 (2):327–345

Gelman A, Carlin JB, Stern HS, Dunson DB, Vehtari A, Rubin DB (2013) Bayesian data analysis. Chapman and Hall/CRC, Stanford, CA

Habier D, Tetens J, Seefried FR, Lichtner P, Thaller G (2010) The impact of genetic relationship information on genomic breeding values in German Holstein cattle. Genet Sel Evol 42(1):5

Habier D, Fernando RL, Kizilkaya K, Garrick DJ (2011) Extension of the Bayesian alphabet for genomic selection. BMC Bioinformatics 12(1):186

Henderson C (1975) Best linear unbiased estimation and prediction under a selection model. Biometrics 31(2):423–447. https://doi.org/10.2307/2529430

Henderson CR, Quaas RL (1976) Multiple trait evaluation using relative's records. J Anim Sci 43: 11–88

Jia Y, Jannink JL (2012) Multiple-trait genomic selection methods increase genetic value prediction accuracy. Genetics 192(4):1513–1522

Jiang J, Zhang Q, Ma L, Li J, Wang Z, Liu JF (2015) Joint prediction of multiple quantitative traits using a Bayesian multivariate antedependence model. Heredity 115(1):29–36

Lehermeier C, Wimmer V, Albrecht T, Auinger HJ, Gianola D, Schmid VJ, Schön CC (2013) Sensitivity to prior specification in Bayesian genome-based prediction models. Stat Appl Genet Mol Biol 12(3):375–391

Meuwissen TH, Hayes BJ, Goddard ME (2001) Prediction of total genetic value using genome-wide dense marker maps. Genetics 157(4):1819–1829

Montesinos-López OA, Montesinos-López A, Crossa J, Toledo FH, Pérez-Hernández O, Eskridge KM, Rutkoski J (2016) A genomic Bayesian multi-trait and multi-environment model. G3 6 (9):2725–2744

Montesinos-López OA, Montesinos-López A, Montesinos-López JC, Crossa J, Luna-Vázquez FJ, Salinas-Ruiz J (2018) A Bayesian multiple-trait and multiple-environment model using the matrix normal distribution. In: Physical methods for stimulation of plant and mushroom development. IntechOpen, Croatia, p 19

Park T, Casella G (2008) The Bayesian lasso. J Am Stat Assoc 103(482):681–686

Pérez P, de los Campos G (2013) BGLR: a statistical package for whole genome regression and prediction. R package version 1(0.2)

Pérez P, de los Campos G (2014) Genome-wide regression and prediction with the BGLR statistical package. Genetics 198(2):483–495

Pérez P, de los Campos G, Crossa J, Gianola D (2010) Genomic-enabled prediction based on molecular markers and pedigree using the Bayesian linear regression package in R. Plant Genome 3(2):106–116

Pollak EJ, Van der Werf J, Quaas RL (1984) Selection bias and multiple trait evaluation. J Dairy Sci 67(7):1590–1595

Pszczola M, Veerkamp RF, De Haas Y, Wall E, Strabel T, Calus MPL (2013) Effect of predictor traits on accuracy of genomic breeding values for feed intake based on a limited cow reference population. Animal 7(11):1759–1768

Schaeffer LR (1984) Sire and cow evaluation under multiple trait models. J Dairy Sci 67 (7):1567–1580

VanRaden PM (2007) Genomic measures of relationship and inbreeding. Interbull Bull 7:33–36

Permissions

All chapters in this book were first published by Springer; hereby published with permission under the Creative Commons Attribution License or equivalent. Every chapter published in this book has been scrutinized by our experts. Their significance has been extensively debated. The topics covered herein carry significant information for a comprehensive understanding. They may even be implemented as practical applications or may be referred to as a beginning point for further studies.

The contributors of this book come from diverse backgrounds, making this book a truly international effort. We would like to thank all the contributing authors for lending their expertise to make the book truly unique. They have played a crucial role in the development of this book. Without their invaluable contributions this book wouldn't have been possible. They have made vital efforts to compile up to date information on the varied aspects of this subject to make this book a valuable addition to the collection of many professionals and students.

This book was conceptualized with the vision of imparting up-to-date and integrated information in this field. To ensure the same, a matchless editorial board was set up. Every individual on the board went through rigorous rounds of assessment to prove their worth. After which they invested a large part of their time researching and compiling the most relevant data for our readers.

The editorial board has been involved in producing this book since its inception. They have spent rigorous hours researching and exploring the diverse topics which have resulted in the successful publishing of this book. They have passed on their knowledge of decades through this book. To expedite this challenging task, the publisher supported the team at every step. A small team of assistant editors was also appointed to further simplify the editing procedure and attain best results for the readers.

Apart from the editorial board, the designing team has also invested a significant amount of their time in understanding the subject and creating the most relevant covers. They scrutinized every image to scout for the most suitable representation of the subject and create an appropriate cover for the book.

The publishing team has been an ardent support to the editorial, designing and production team. Their endless efforts to recruit the best for this project, has resulted in the accomplishment of this book. They are a veteran in the field of academics and their pool of knowledge is as vast as their experience in printing. Their expertise and guidance has proved useful at every step. Their uncompromising quality standards have made this book an exceptional effort. Their encouragement from time to time has been an inspiration for everyone.

The publisher and the editorial board hope that this book will prove to be a valuable piece of knowledge for students, practitioners and scholars across the globe.

Index

Printed in the USA
CPSIA information can be obtained
at www.ICGtesting.com
LVHW010814221023
761596LV00005B/492

9 798888 360521